I0810275

POSTCARD HISTORY SERIES

Western Pennsylvania's Oil Heritage

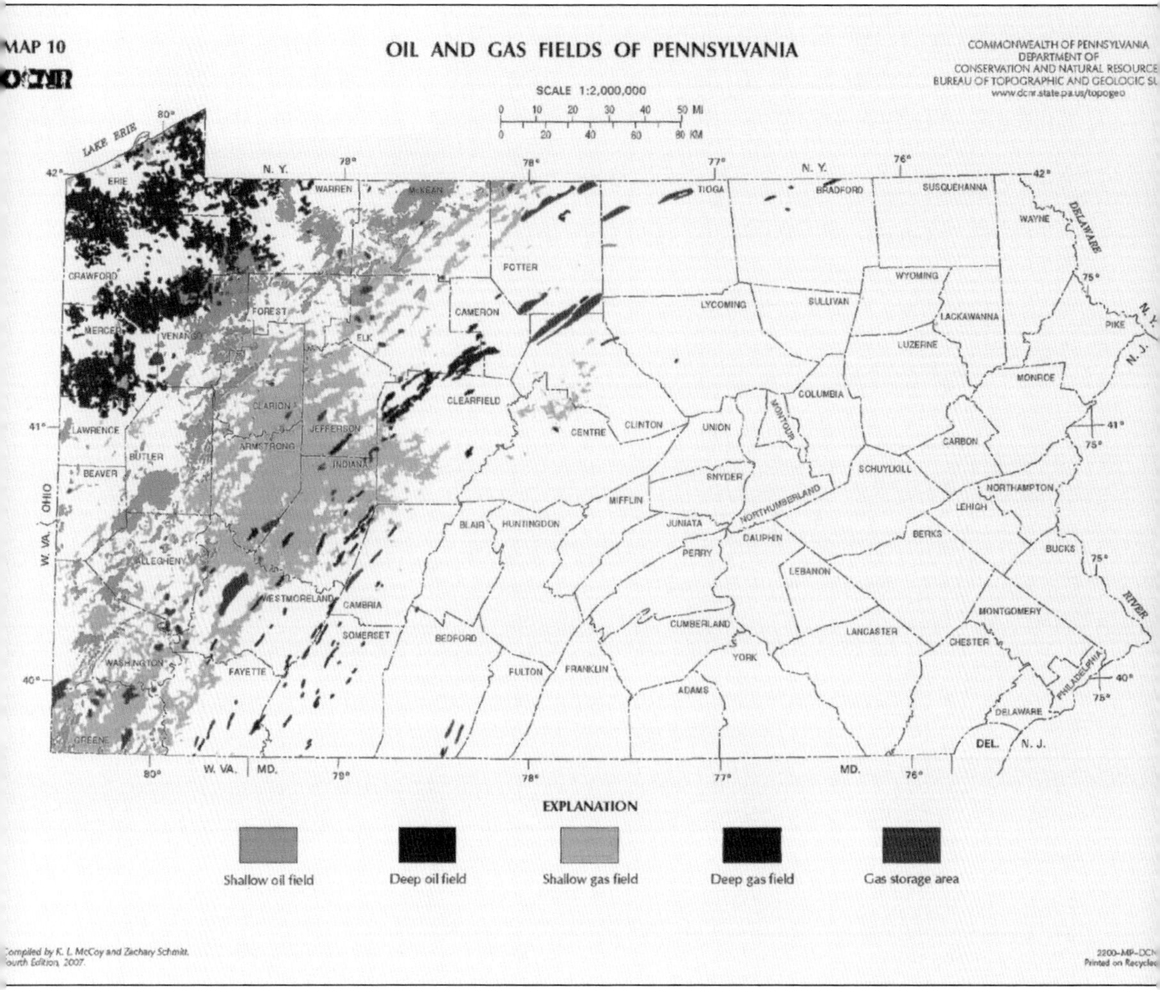

The map shows the distribution of oil and gas in Pennsylvania. (Courtesy of Commonwealth of Pennsylvania, Department of Conservation and Natural Resources, Bureau of Topographic and Geologic Survey, 2007.)

On the front cover: Please see page 30. (Author's collection.)

On the back cover: Please see page 20. (Author's collection.)

POSTCARD HISTORY SERIES

Western Pennsylvania's Oil Heritage

Charles E. Williams

ARCADIA PUBLISHING

ISBN 978-0-7385-6304-6

Published by Arcadia Publishing
Charleston, South Carolina

Printed in the United States of America

Library of Congress Catalog Card Number: 2008924574

For all general information contact Arcadia Publishing at:
Telephone 843-853-2070
Fax 843-853-0044
E-mail sales@arcadiapublishing.com
For customer service and orders:
Toll-Free 1-888-313-2665

Visit us on the Internet at www.arcadiapublishing.com

Dedicated to Kim, Tara, Kelsey, and Owen

Contents

ACKNOWLEDGMENTS

I thank my family, friends, and colleagues who have supported my work on the history of northern Pennsylvania landscapes over the years. I especially give heartfelt thanks to my wife, Kim, and our daughters, Tara and Kelsey. Their support, encouragement, and understanding made this book possible.

Finally, I extend my thanks to the folks at Arcadia Publishing, especially Erin Vosgien, junior publisher, for their patience, encouragement, and support.

Introduction

As a young student in history class, I learned about the successful drilling of the first oil well in 1859 by Col. Edwin Drake and his crew on Oil Creek in western Pennsylvania. Having no apparent impact other than helping to prepare me for exam questions, this knowledge was buried in my mind for nearly two decades. That is, until I moved to western Pennsylvania's oil region.

The oil boom had passed well over a century before I began exploring the landscapes of northwestern Pennsylvania as part of my research in regional forest ecology. I was stunned by how much historical infrastructure still remained in the regenerating forest: well casings, pipelines, old pump jacks, and defunct powerhouses. If you were not alert, you could trip on these artifacts in areas where ferns carpeted the forest floor.

At first, I was taken aback by the widespread industrial footprint that perforated the forest and peppered some of my study sites in the oil region. I had seen none of this type of extensive industrial infrastructure in my earlier work in forests of the southern Appalachian Mountains, where oil development was largely absent. I then began to think about how the historic oil boom had shaped not just the local forests that I studied, but also the regional culture, economy, and larger landscape of western Pennsylvania's oil region. The artifacts and historical infrastructure of the oil boom that I found on the landscape became of growing interest, and I slowly started to remember, and add to, my old history lessons from decades ago. The western Pennsylvania oil region was a cultural landscape—contemporary forests influenced by oil history—worthy of further study.

Few discoveries have shaken the world as much as that of oil flowing from the wilds of northwestern Pennsylvania in 1859. Known since antiquity in many cultures as a fuel, a medicine, and an anointment, oil now emerged on the world scene in industrial fashion. Refined as kerosene for lamps, a lubricant for home and industrial needs, and ultimately as a fuel for transportation in the form of gasoline, oil literally changed the world and defined global economics and politics for much of the 20th century. Globally it is likely that few people appreciate that the oil industry started in the oil fields of western Pennsylvania a century and a half ago.

Images of oil gushers in Texas, like those at Spindletop, often come to mind when one thinks about oil prospecting and the development of the industry. But it is the dogged determination of early Pennsylvania oilmen like Colonel Drake, who helped drill the first successful well near Titusville along Oil Creek, and those that followed him, that kick-started the age of oil. Oil production in the United States shifted from Pennsylvania to other states beginning in the late 1800s, when new oil fields were discovered. Pennsylvania, once the nation's leading oil producer, continues to produce high-quality oil for regional, national, and international markets to this date.

Efforts to preserve the history of oil in western Pennsylvania have grown considerably as we approach the 150-year anniversary of Colonel Drake's success. Drake Well State Park near Titusville is a prime example, with a museum, exhibits, and an extensive library on oil history. The adjacent Oil Creek State Park preserves and interprets portions of the oil boom landscape, including the ghost town of Petroleum Centre and several historic "oil farms" such as the Benninghoff Farm. The recent designation of portions of western Pennsylvania's oil region as both state and national heritage regions has also done much to promote the appreciation of oil history. Many individuals have also left an important stamp on communicating the oil region's history. Perhaps the most notable person in this regard is the photographer John A. Mather, who lived in Titusville. Mather roamed the oil region during the boom times of the 1860s by wagon and even traversed Oil Creek with his own floating office. His images of the early oil landscapes, thousands of which are deposited at Drake Well State Park's library, are priceless reminders of the region's past. Some of his images, transformed into postcards, appear in this book.

The images of the western Pennsylvania oil region, on which this book is based, come from a variety of postcards, largely from the early 1900s. My interest in writing this book was not to provide an in-depth history of western Pennsylvania's oil heritage. Instead I provide vignettes of the oil region's past via postcard images that chronicle the landscape, its key towns and cities, and its people. I have used a geographical approach to order the content of this book as I have done with my previous postcard history books of Pennsylvania's Allegheny River region. I hope you enjoy the tour.

One

Oil Creek and the Early Days of the Oil Industry

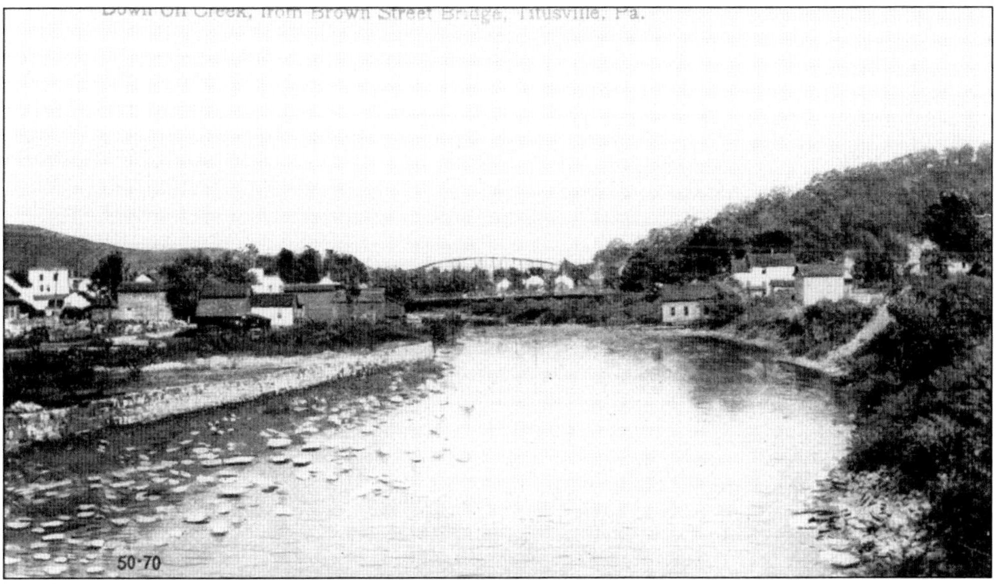

Oil Creek, a tributary of the Allegheny River, was long known to be a source of Seneca or "snake oil" that was often used or sold locally in small quantities as a folk medicine. Oil was collected by scooping it in wooden bowls or by soaking it up in rags. This tranquil image of Oil Creek near Titusville in the early 1900s belies the frenzy and chaos that followed the world's first oil boom in 1859.

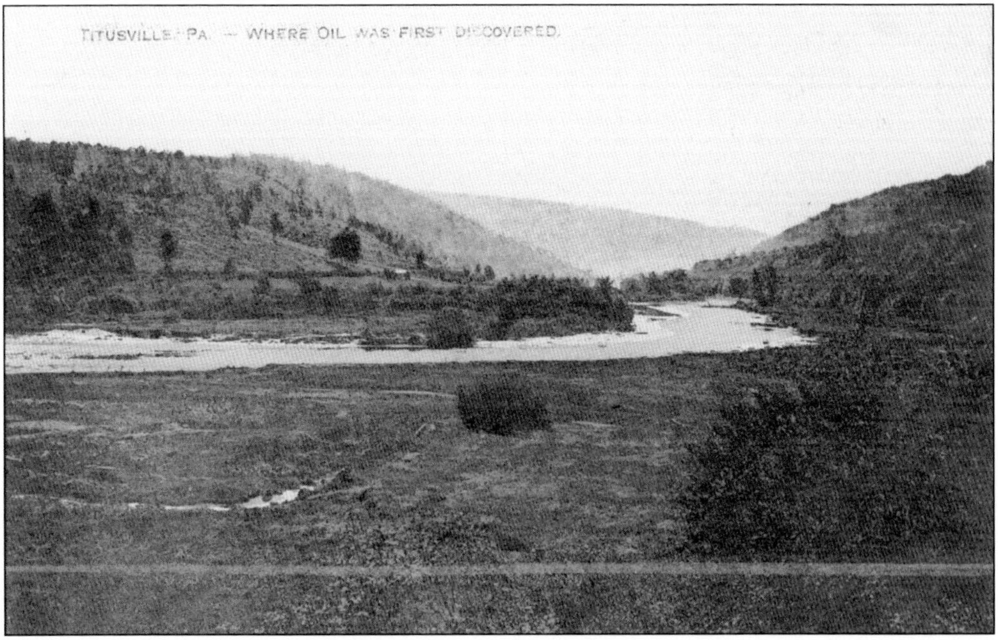

Oil Creek is located in Erie, Crawford, Venango, and Warren Counties and is formed from the union of its east and west branches at Centerville, Crawford County. The Oil Creek Valley broadens to nearly a mile in width below Titusville and was a landscape of forest and small farms before the oil boom. This image of Oil Creek from the early 1900s shows the landscape recovering from the boom.

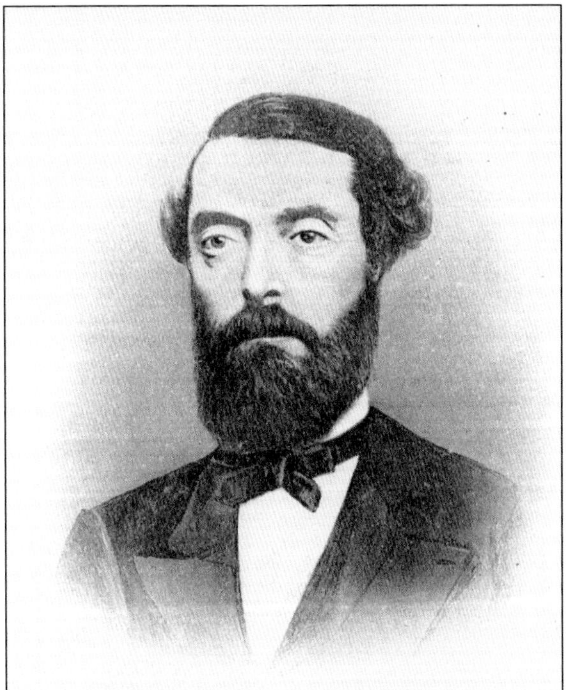

51297 Col. Drake, Driller of the First Oil Well

The commercial potential of Oil Creek crude was first recognized by George H. Bissell and Jonathan G. Eveleth, who organized the Pennsylvania Rock Oil Company in 1854. Edwin L. Drake, a former railroad employee, was sent to Titusville to examine the oil farm that the company had obtained along Oil Creek. Drake's report was favorable, and a new company was organized to explore for oil with Drake as an agent—the Seneca Oil Company.

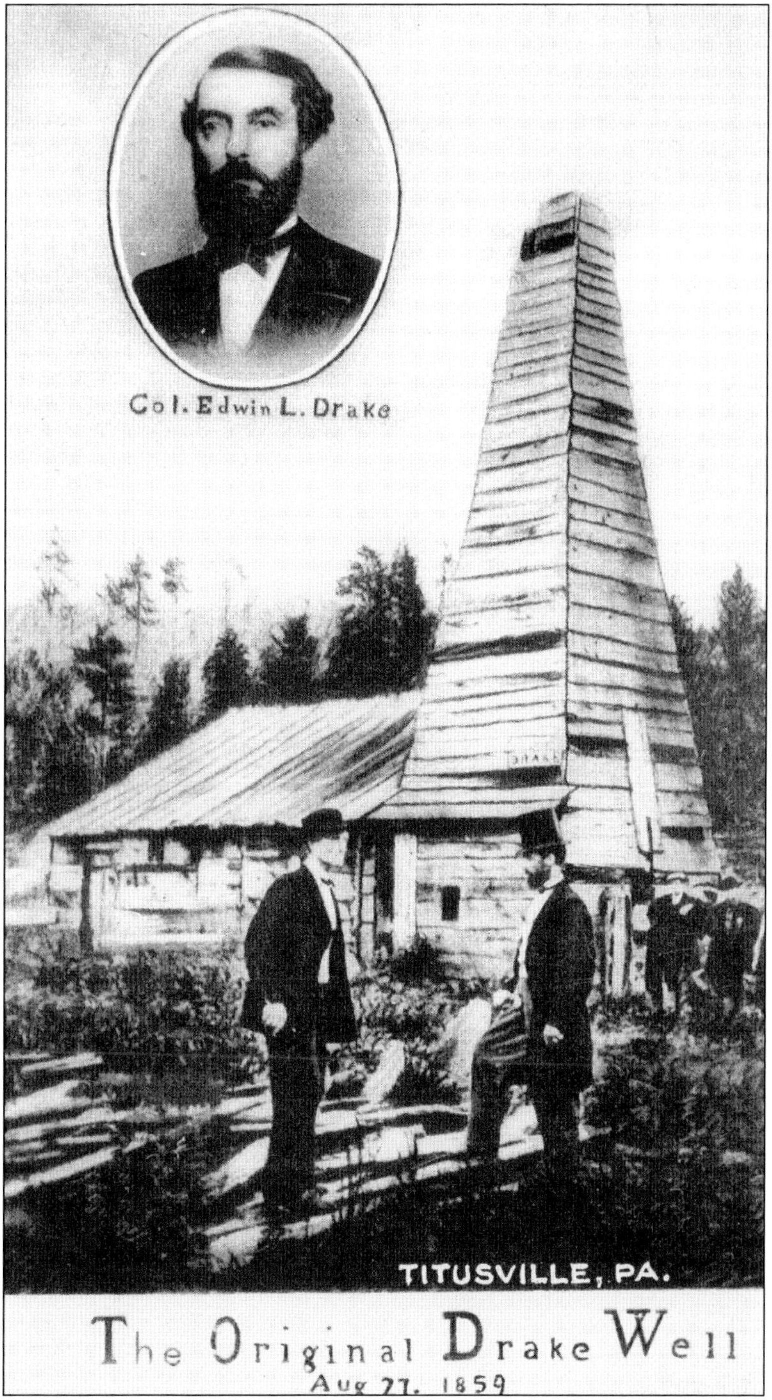

Drake was convinced that oil could be raised from the ground in the same manner as salt was in western Pennsylvania, by drilling a well. After going through several salt well drillers, Drake finally hired William A. "Uncle Billy" Smith, a blacksmith and salt well driller from Tarentum, to drill for oil. On August 27, 1859, they were successful, and the oil boom began. This image from 1861 shows Drake on the right with his friend Peter Wilson, a Titusville druggist, on the left.

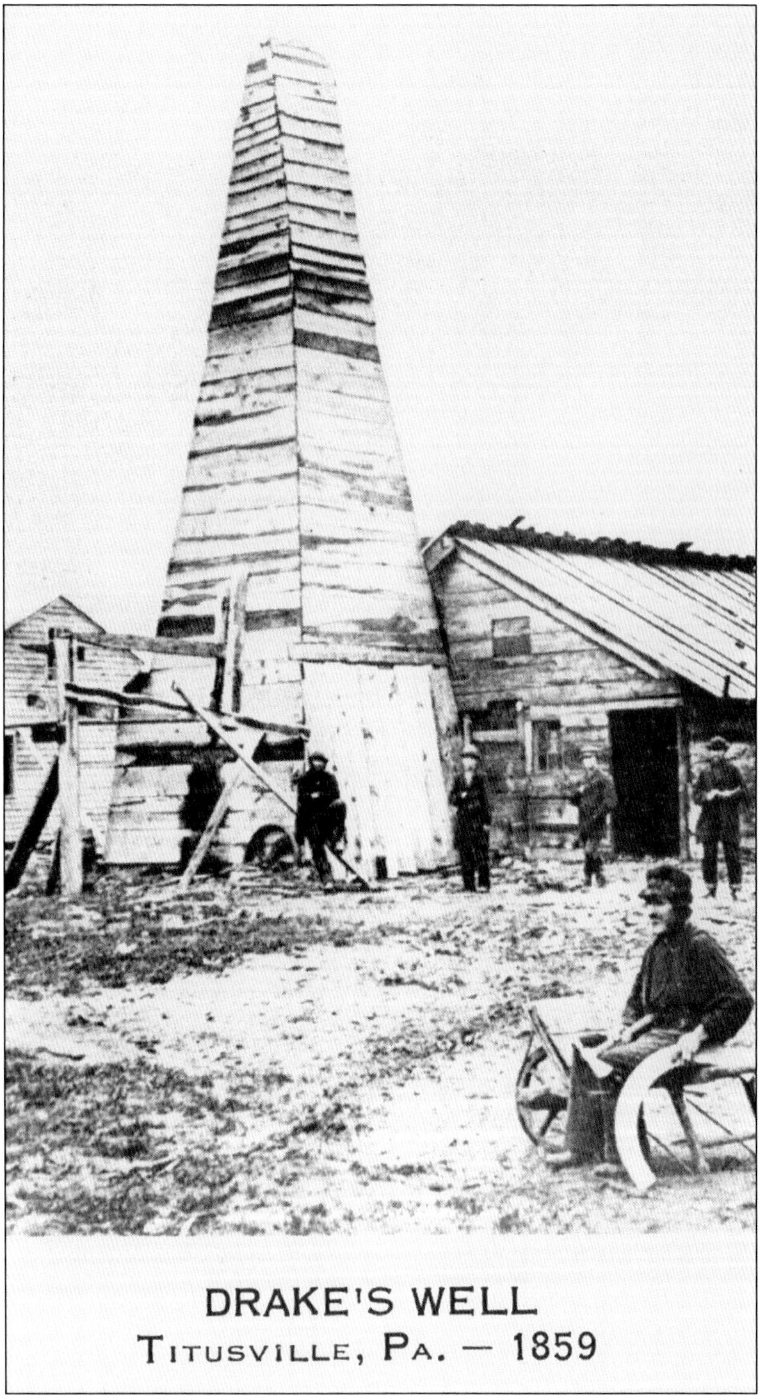

This image shows Drake Well in 1864. Uncle Billy Smith sits on the wheelbarrow at the image's bottom right. Both this image and that on the opposite page show the second building and derrick at the well, the original ones having been destroyed by a fire that Smith accidentally started in 1859. Smith was paid $2.50 a day for his work with Drake, which included the services of his son Samuel.

After Edwin Drake's success, oil well drilling evolved rapidly. Initially the spring pole method, in which two men "kicked down a well" with a wooden fulcrum attached to drilling tools, was the chief process. Drake employed a steam engine to power his drilling equipment and struck oil at a depth of 69.5 feet. This image shows the inside of a derrick as it may have appeared during Drake's time.

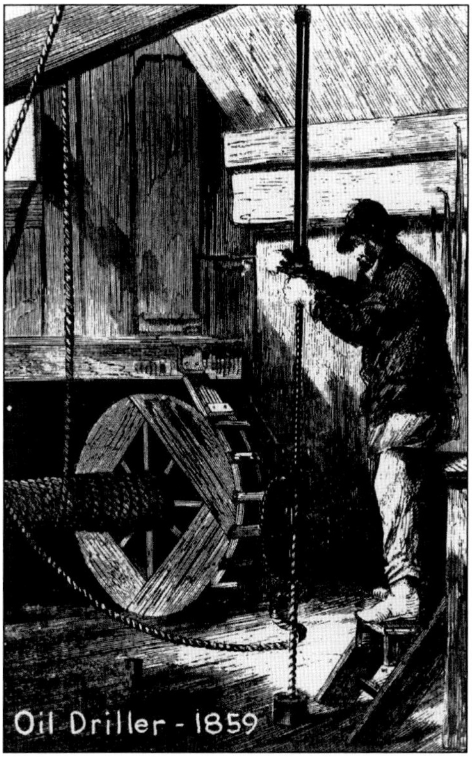

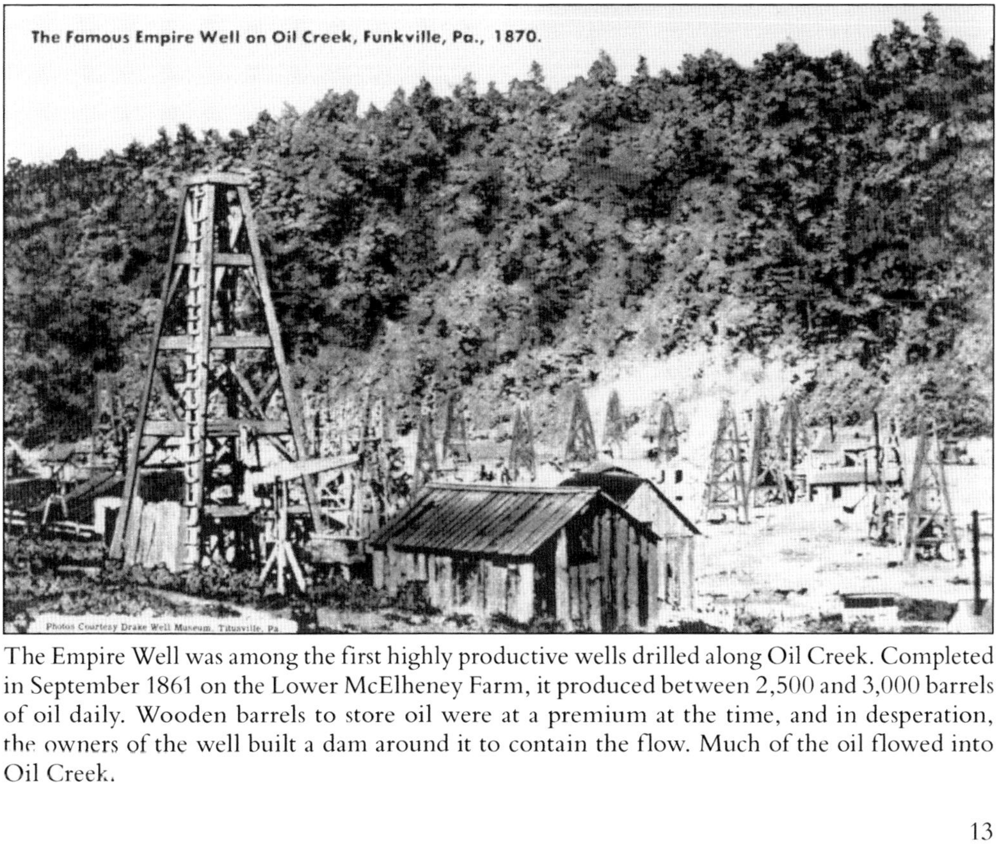

The Empire Well was among the first highly productive wells drilled along Oil Creek. Completed in September 1861 on the Lower McElheney Farm, it produced between 2,500 and 3,000 barrels of oil daily. Wooden barrels to store oil were at a premium at the time, and in desperation, the owners of the well built a dam around it to contain the flow. Much of the oil flowed into Oil Creek.

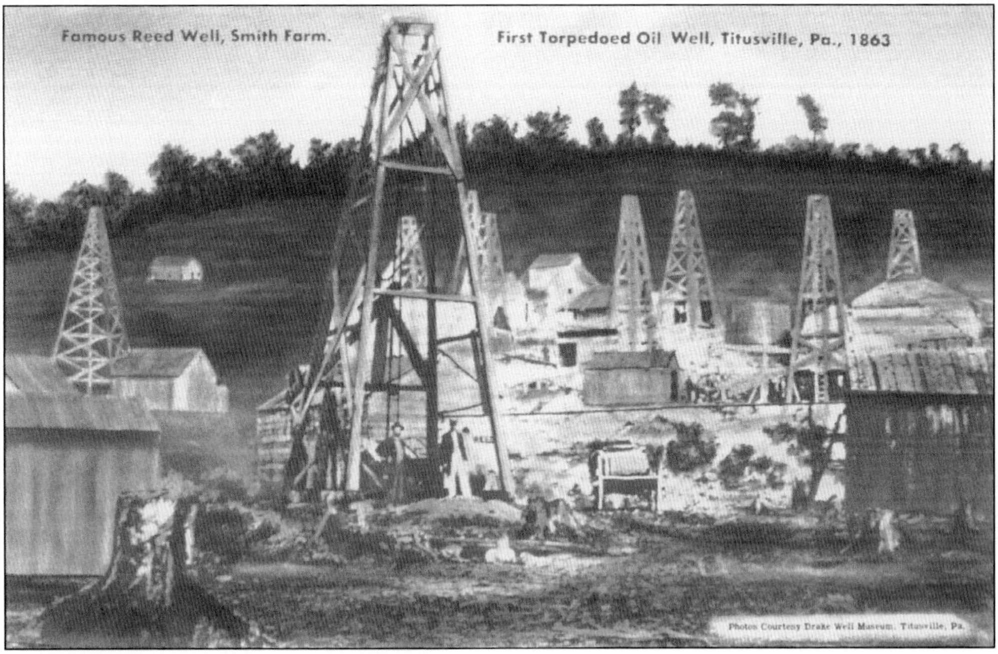

The Smith Farm along Cherry Run, a tributary of Oil Creek near Rouseville, was first drilled by William Reed. The Reed Well, shown in this image, struck oil at a depth of 610 feet and produced between 200 and 300 barrels daily. William Reed claimed to be the inventor of the torpedo, a gunpowder-laden missile used to stimulate oil flow in wells. His experiments in 1863 on the Reed Well were considered failures.

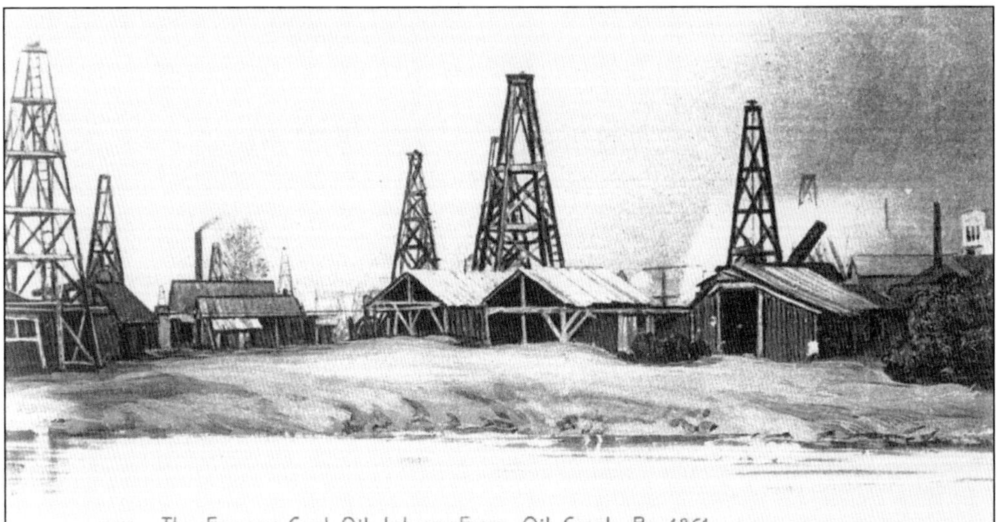

The Culbertson McClintock farm was located on the west bank of Oil Creek, four miles north of Oil City. The farm came into production in 1861 to 1862. Early oil production from this farm was estimated at 350 barrels a day. Culbertson McClintock died before oil was struck on his farm, but his wife, Sarah, became wealthy from oil leases. She was fatally burned in March 1864 when she threw crude oil on her kitchen stove to hurry the fire.

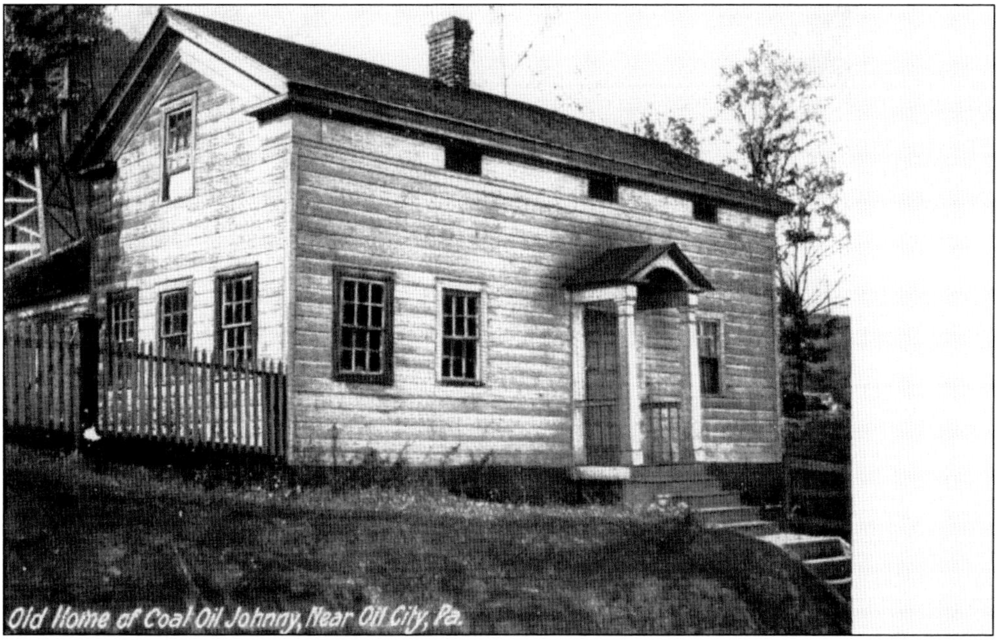

"Coal Oil Johnny," or John W. Steele, was a well-known figure from the Pennsylvania oil region. In 1864, at 20 years of age, Steele inherited the McClintock farm, the house shown above, and the farm's lucrative wells on the death of his adoptive mother Sarah. The wells gave him a daily income of $2,000. The nickname "Coal Oil Johnny" was given to Steele by reporters who followed the excesses of his later life, which ended in bankruptcy.

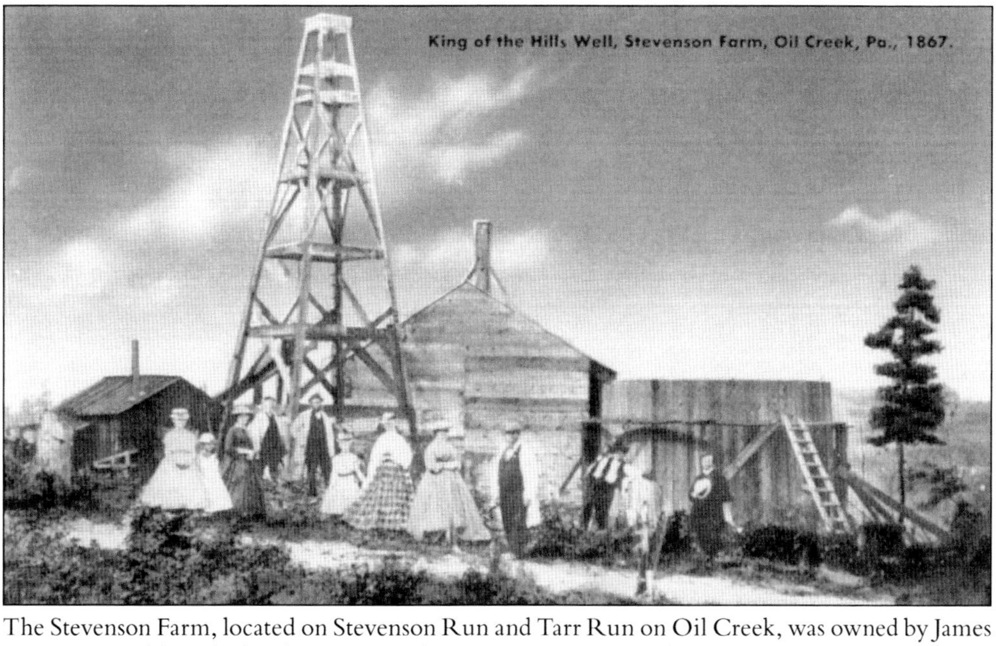

The Stevenson Farm, located on Stevenson Run and Tarr Run on Oil Creek, was owned by James Stevenson and bought by the Ocean Oil Company in 1864. It began producing in September 1865. By 1870, nine producing oil wells were on the property with an average daily production of 1,000 barrels. The King of the Hills Well, shown in this image, was completed in June 1866 and produced about 300 barrels daily.

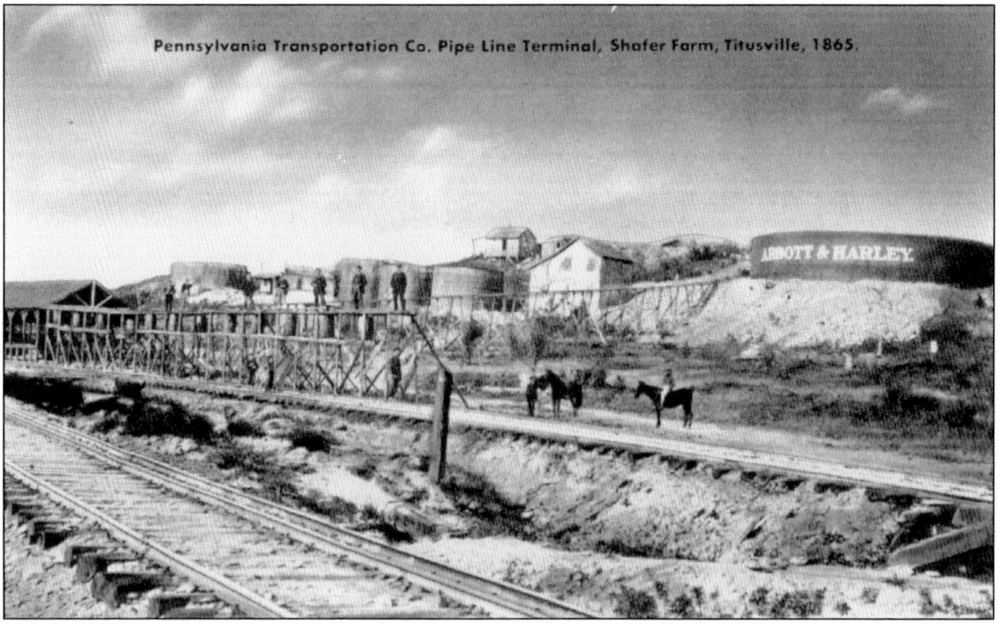

The Shaffer Farm was located six miles south of Titusville on both sides of Oil Creek. Not known for its oil production, Shaffer Farm became an important site for the staging and transportation of oil. The Pennsylvania Transportation Company, which controlled over 500 miles of oil pipeline by the 1870s, had a terminal at Shaffer Farm. At the time, Shaffer Farm was also the southern terminus of the Oil Creek Railroad.

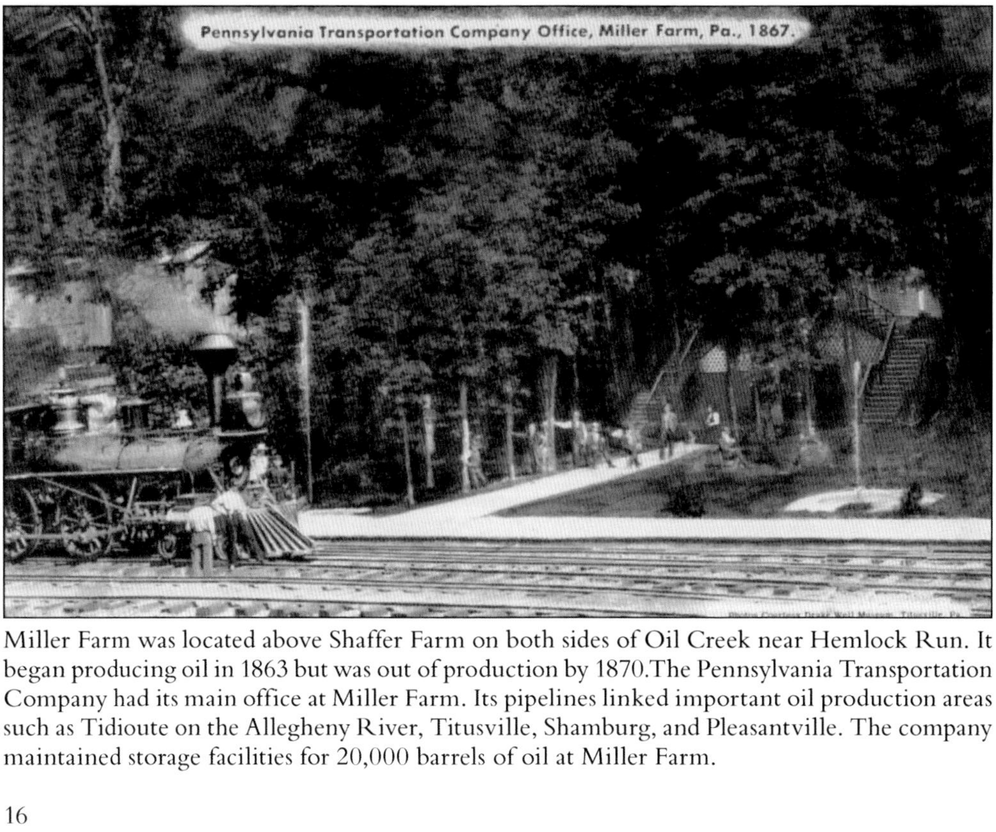

Miller Farm was located above Shaffer Farm on both sides of Oil Creek near Hemlock Run. It began producing oil in 1863 but was out of production by 1870.The Pennsylvania Transportation Company had its main office at Miller Farm. Its pipelines linked important oil production areas such as Tidioute on the Allegheny River, Titusville, Shamburg, and Pleasantville. The company maintained storage facilities for 20,000 barrels of oil at Miller Farm.

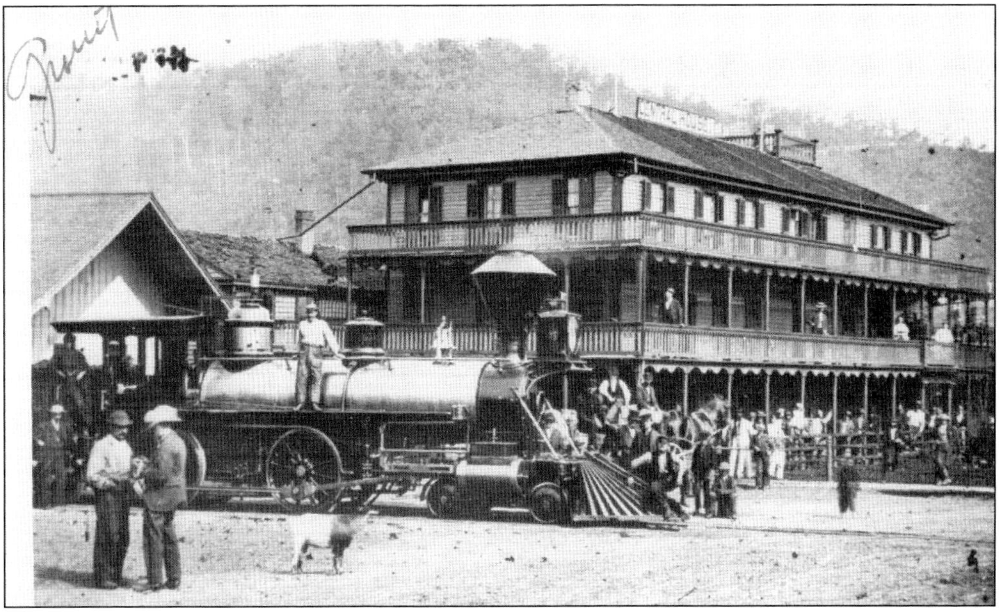

Petroleum Centre was established in 1864 by the Central Petroleum Company of New York and earned its name by being halfway between Titusville and Oil City. It was one of the early boomtowns of the oil region, quickly attracting 3,000 people—sprouting hotels, liveries, saloons, gambling halls, and banks virtually overnight. This image shows the Oil Creek Railroad's locomotive at Petroleum Centre, which reached the town in 1866.

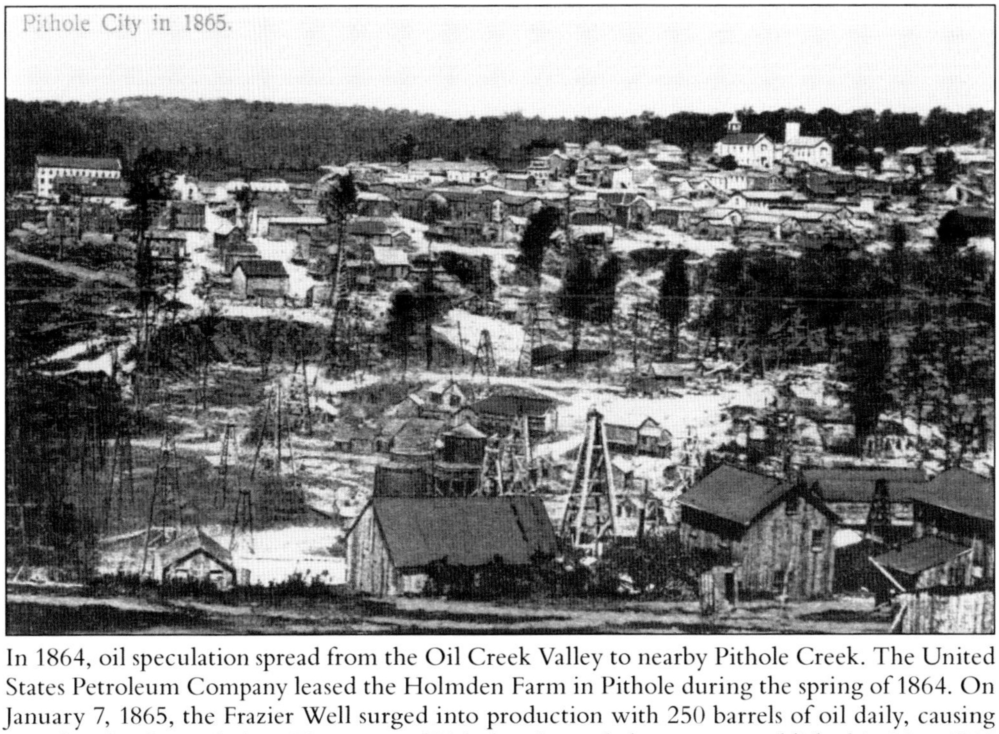

In 1864, oil speculation spread from the Oil Creek Valley to nearby Pithole Creek. The United States Petroleum Company leased the Holmden Farm in Pithole during the spring of 1864. On January 7, 1865, the Frazier Well surged into production with 250 barrels of oil daily, causing a mad rush of speculation. The town of Pithole, pictured above, was established in May 1864 and had a population of 15,000 by September.

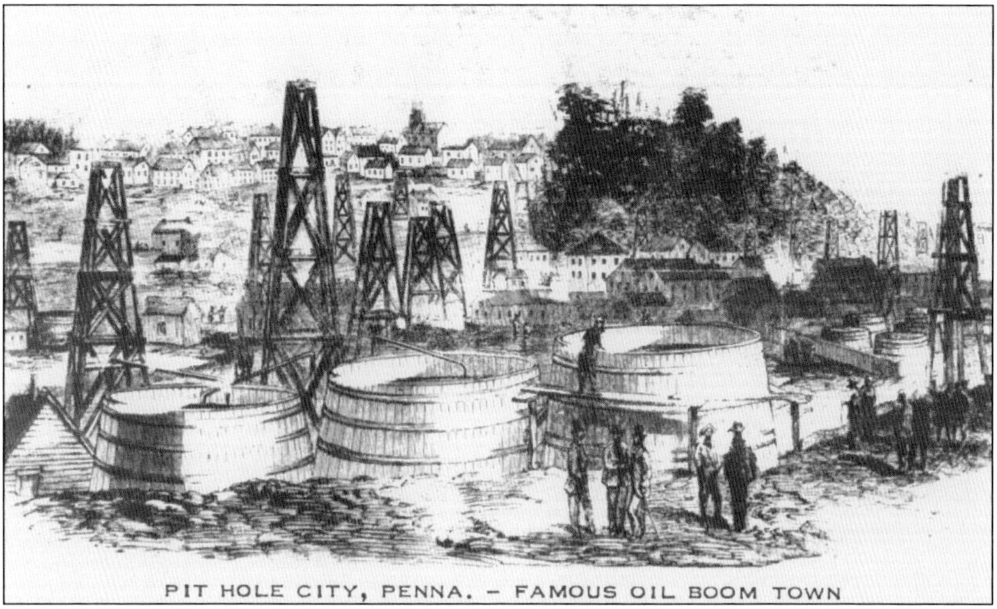

Drilling of new wells proceeded rapidly along Pithole Creek in 1865. The Homestead Well was the second oil well drilled at Pithole and came into production in April 1865 at 50 barrels daily. It would later produce 500 barrels daily. The Grant Well began flowing in August 1865 at 200 barrels daily but eventually reached 1,200 barrels daily. Besides derricks, large wooden storage tanks, like those seen in the image above, dominated the landscape around Pithole.

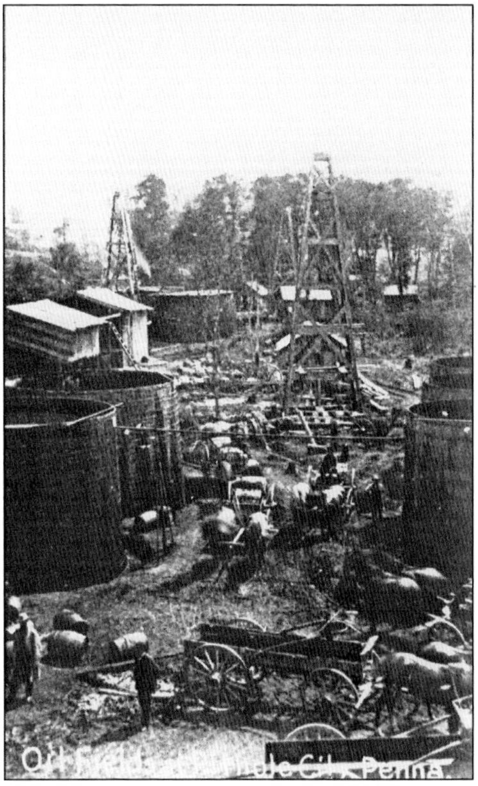

There were no pipelines through which to transport the abundant oil flowing from Pithole until October 1865. Until then, transportation of oil was done by horse and oxen teams. Teamsters, who often charged exorbitant prices, were in heavy demand, along with wood for building storage tanks and barrels for transport. This image taken in 1865 shows a frenzy of teams queuing up for oil barrels at the Frazier Well at Pithole.

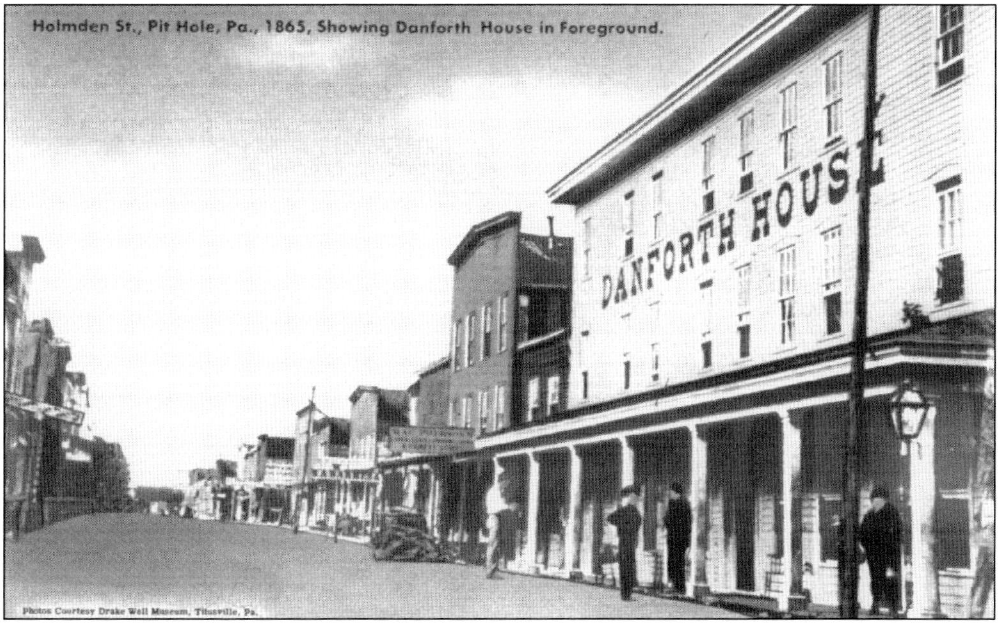

By September 1865, Pithole boasted two banks, two telegraphs, over 50 hotels, a daily newspaper, and one of the largest post offices in Pennsylvania, handling 3,000 pieces of mail daily. This image of Pithole from 1865 shows Holmden Street and the Danforth House. The Danforth House, built at a cost of $40,000, opened on Christmas Day 1865 and could accommodate 140 guests. One of the last buildings at Pithole, it was sold for firewood in 1885 for $16.

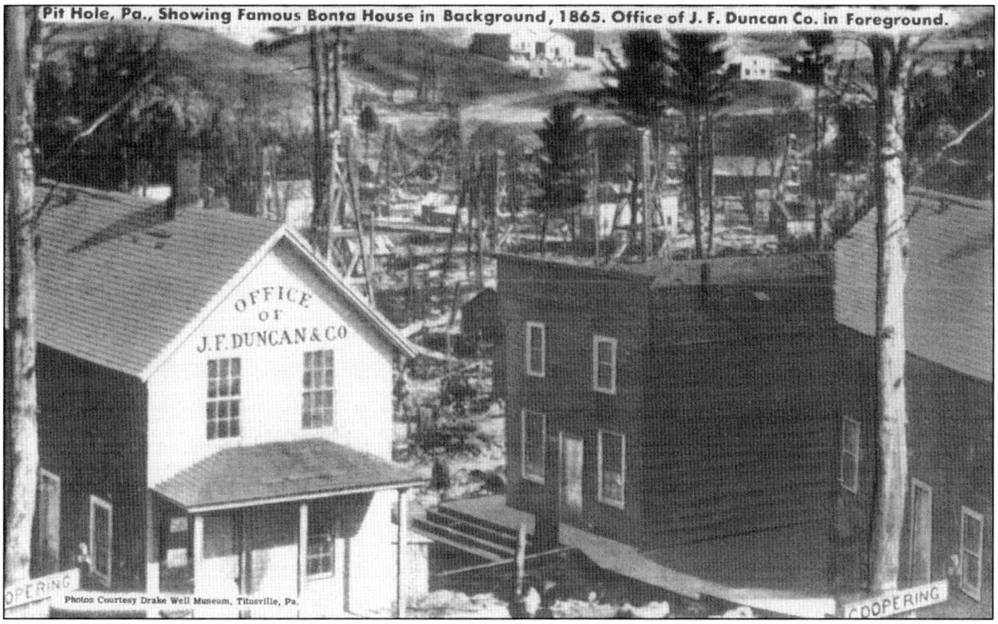

This image of Pithole in 1865 shows the office of the J. F. Duncan Company with the Bonta House hotel in the distance. The Bonta House was built in 1865 at a cost of $80,000 and was probably the finest hotel in Pithole with rooms for 175 guests. The Bonta House was eventually sold to Franklin S. Tarbell, father of "muckraker" journalist Ida Tarbell, for $600. Tarbell disassembled the hotel and used the materials to build his home in Titusville.

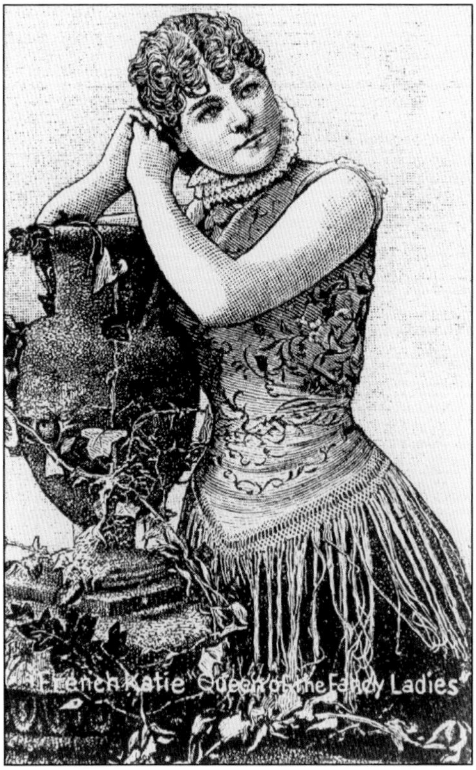

Pithole had a reputation for lawlessness and attracted a wide range of people. Among them was Ben Hogan, the self-proclaimed "wickedest man in the world" who opened a "sporting palace" at Pithole. French Kate, seen at left, one of Hogan's associates, was considered one of the "fast ladies" of the town. Like its rapid rise, Pithole quickly crashed when oil ceased to flow. By January 1866, the throng had left, and Pithole was deserted.

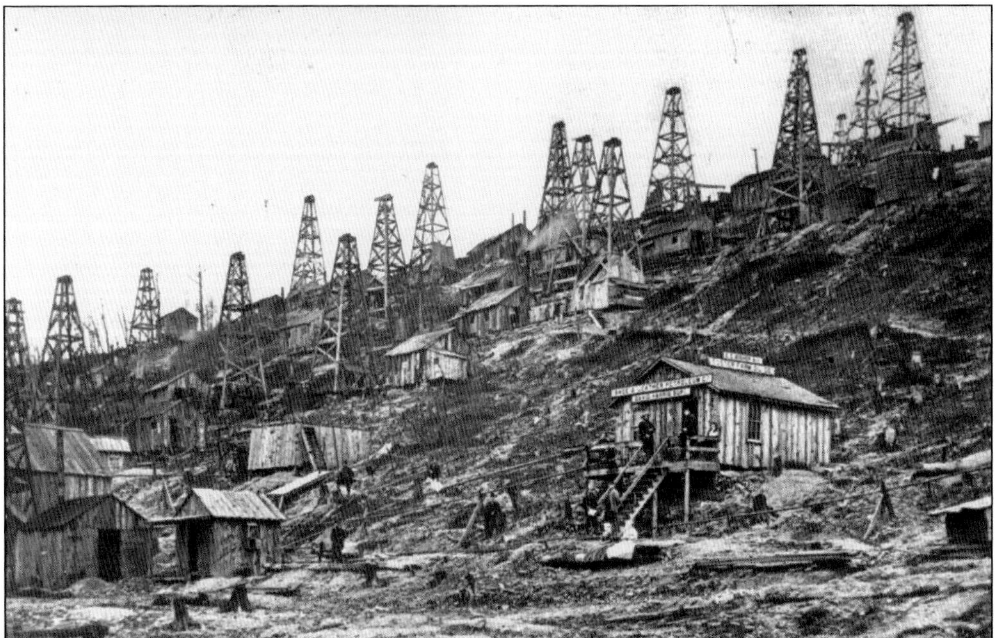

Pioneer was an oil boomtown that grew up at the confluence of Oil Creek and Pioneer Run in the mid-1860s. Speculation along Pioneer Run was bolstered by the successful Lady Brooks Well, opened in 1866, which produced about 600 barrels of oil daily. This image, taken by John Mather in 1865, shows the intensity of development along Pioneer Run.

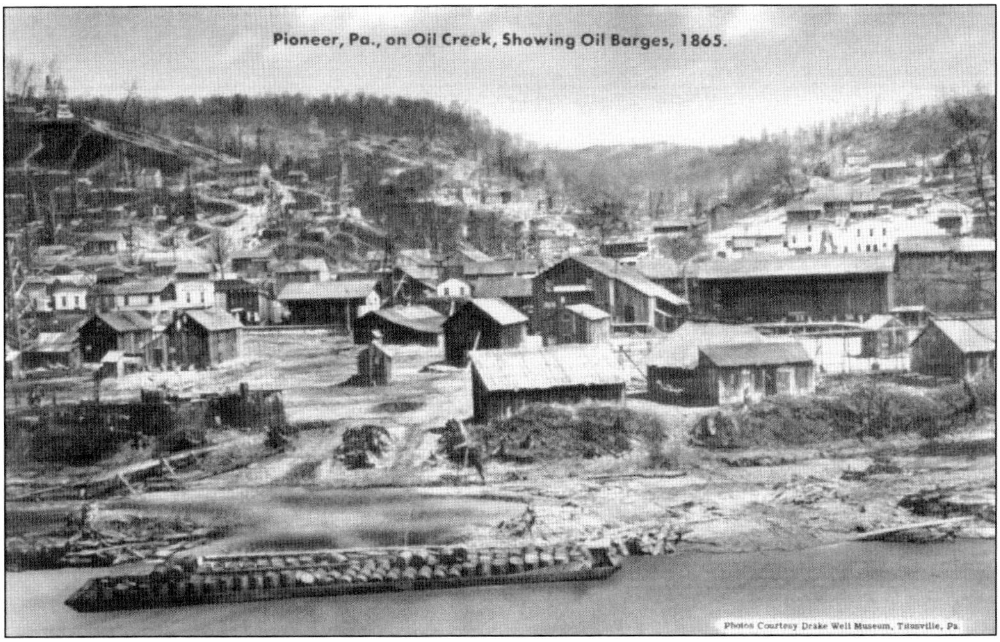

Pioneer was also an important oil shipping and staging area on Oil Creek. Barrels were critical to shipping oil prior to the advent of pipelines, as can be seen in the stack of barrels along Oil Creek at the bottom of this image. Supply rarely met demand, and barrel prices were often at a premium. White oak was in high demand for barrels because its wood was less porous than that of other species and held oil better.

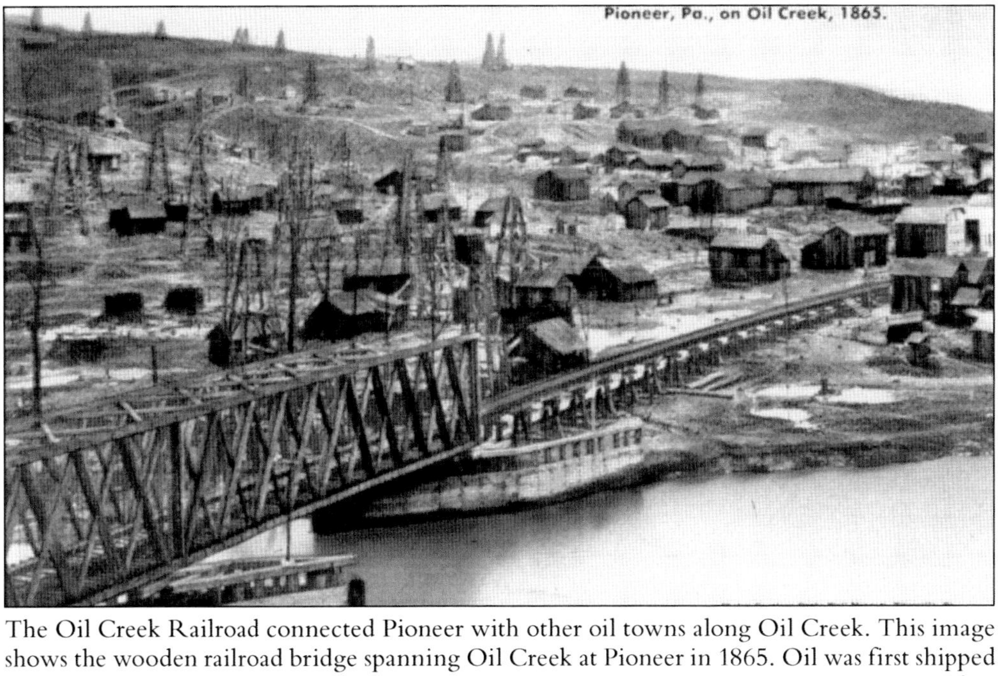

The Oil Creek Railroad connected Pioneer with other oil towns along Oil Creek. This image shows the wooden railroad bridge spanning Oil Creek at Pioneer in 1865. Oil was first shipped on standing barrels in railroad flatcars. Later Amos and James Densmore of Miller Farm modified a flatcar to carry two wooden tanks capable of holding 40 to 45 barrels of oil. Thus the railroad "tank car" was born.

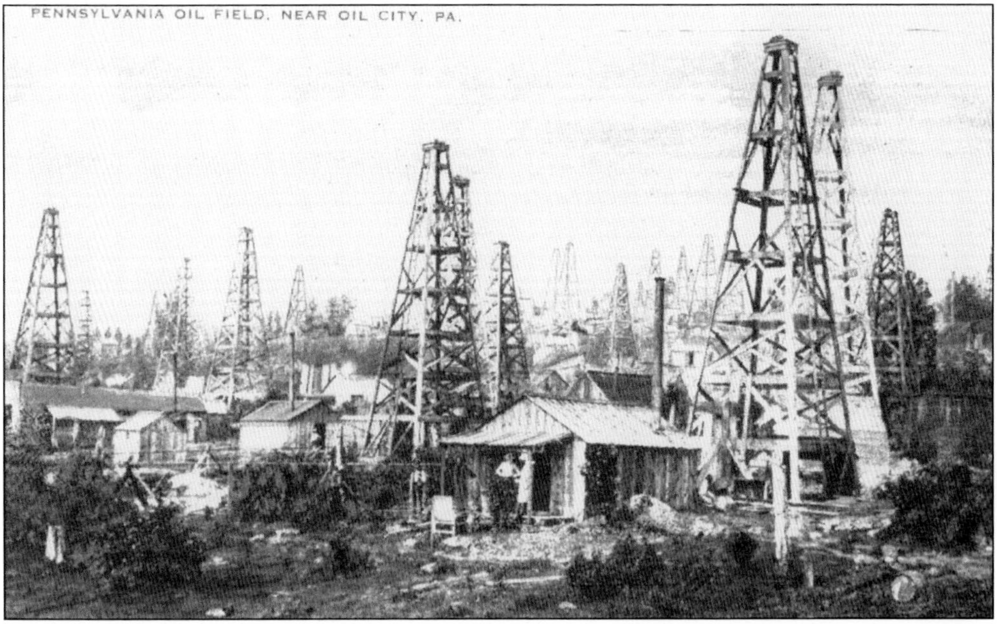

Speculators began to drill for oil along Cherrytree Run, a tributary of Oil Creek, in the late 1860s. Eventually over 300 wells would be developed along this stream, although few were prolific producers. Among these was the oddly named Allemagoozelum Well, drilled along Cherrytree Run in the late 1860s. This image shows the Hess Farm Wells along Cherrytree Run in 1869.

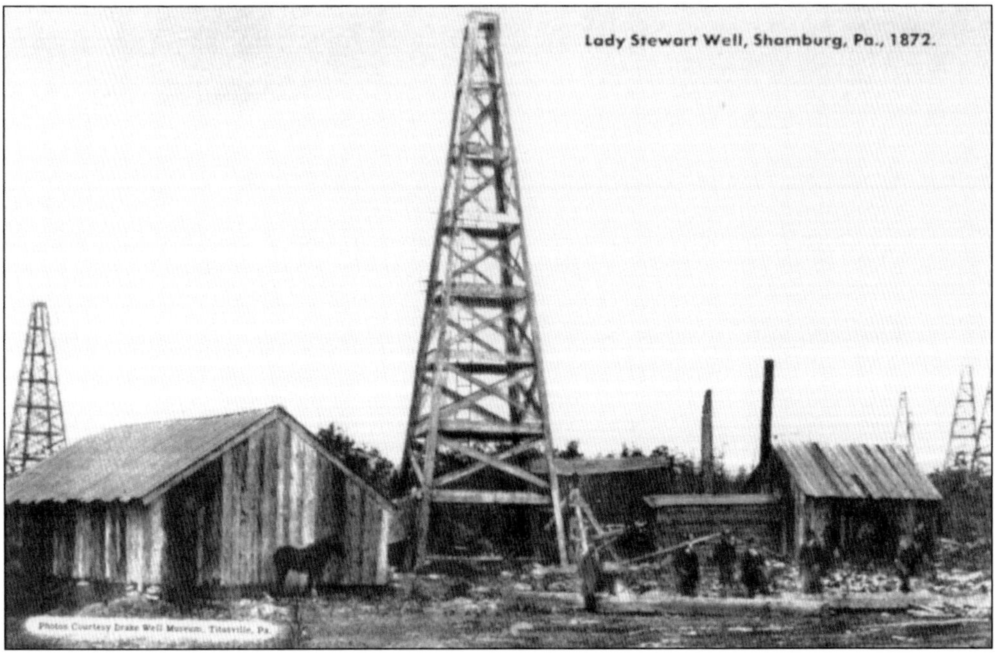

Shamburg got its start in 1866 along the upper headwaters of Cherry Run, a tributary of Oil Creek in northern Venango County. It was named after Dr. G. S. Shamburg, an early resident and oil developer. At its boom in 1868, Shamburg had 4,000 residents. Drillers in the Shamburg oil pool purportedly used the best drilling tools and built among the tallest derricks of the time. The Lady Stewart Well, shown here, produced 180 barrels daily in 1869.

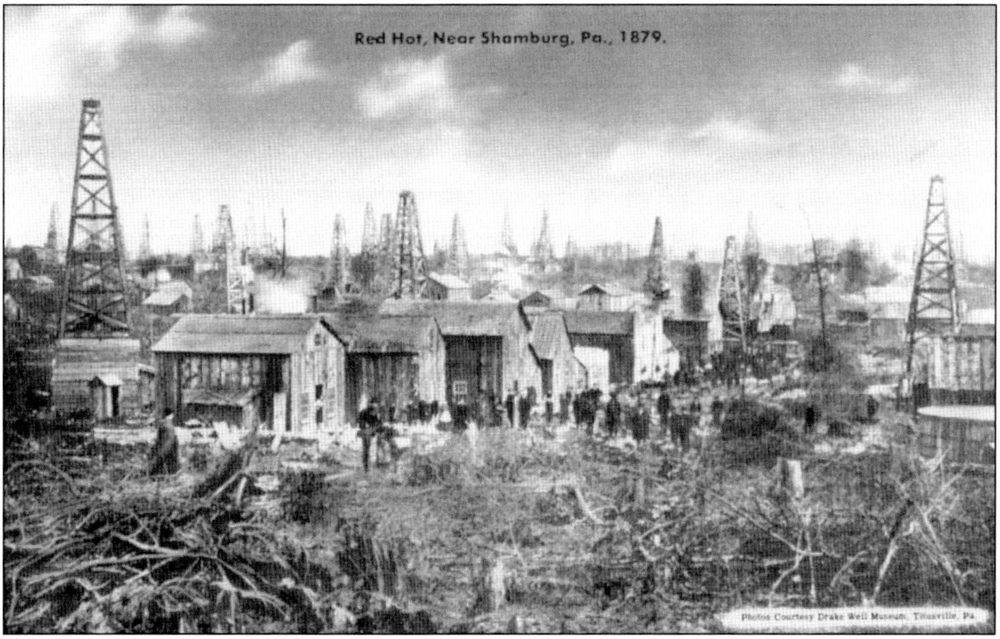

Red Hot sprang up near Shamburg in 1869, after a well drilled there by the Windsor Brothers of Oil City yielded 500 barrels of oil daily. Red Hot, a collection of derricks and cabins, showed little evidence of permanency, the inhabitants largely using the services and supplies afforded by nearby Shamburg. This image of Red Hot was taken in 1870, a year before the wells played out and the oilmen moved on.

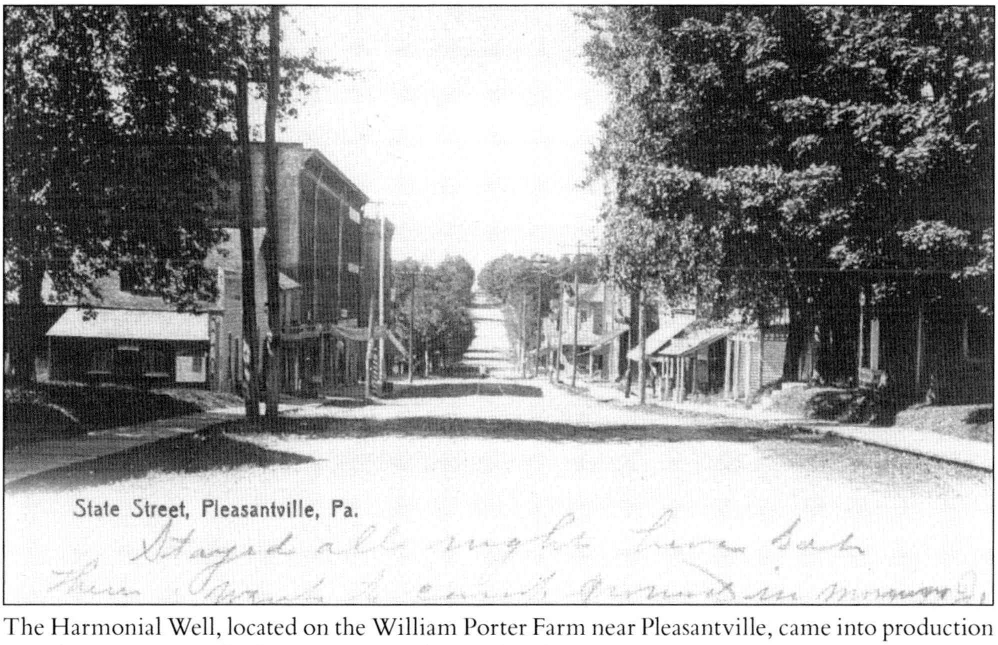

The Harmonial Well, located on the William Porter Farm near Pleasantville, came into production in February 1868 and ushered in a short boom for the quiet town. By the summer of 1868, the population of Pleasantville had tripled from 1,000 to 3,000. An established town before the oil boom, Pleasantville generally lacked the lawlessness and mayhem of boomtowns that sprang up from nothing overnight. This view of State Street, Pleasantville, was postmarked in 1907.

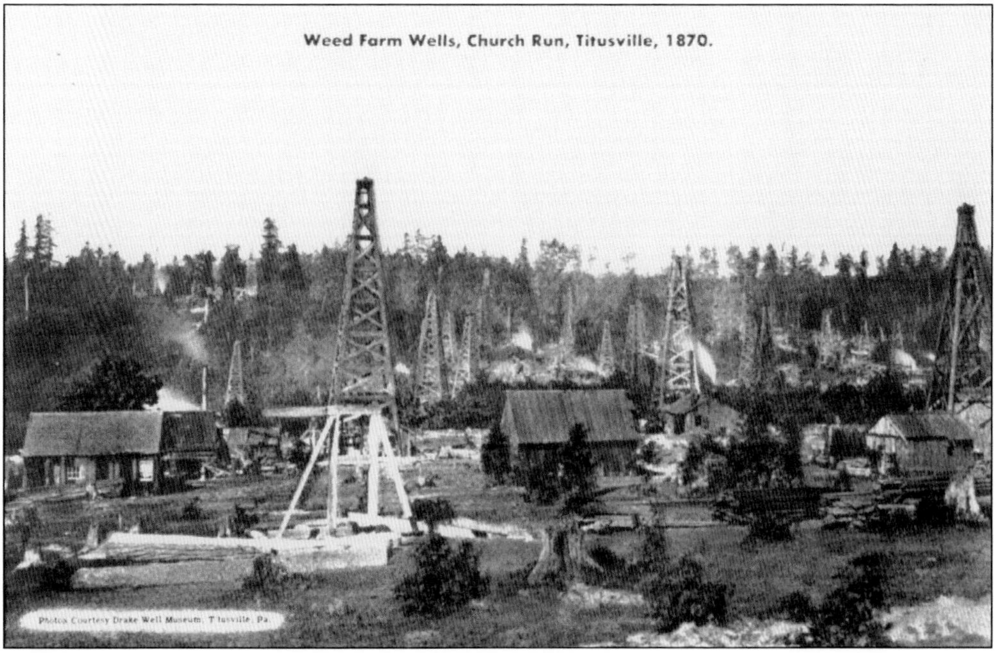

Church Run near Titusville was not extensively explored for oil until 1870. A good reliable product was found along Church Run, although individual well production was not large. This image shows the Weed Farm Wells on Church Run in 1870.

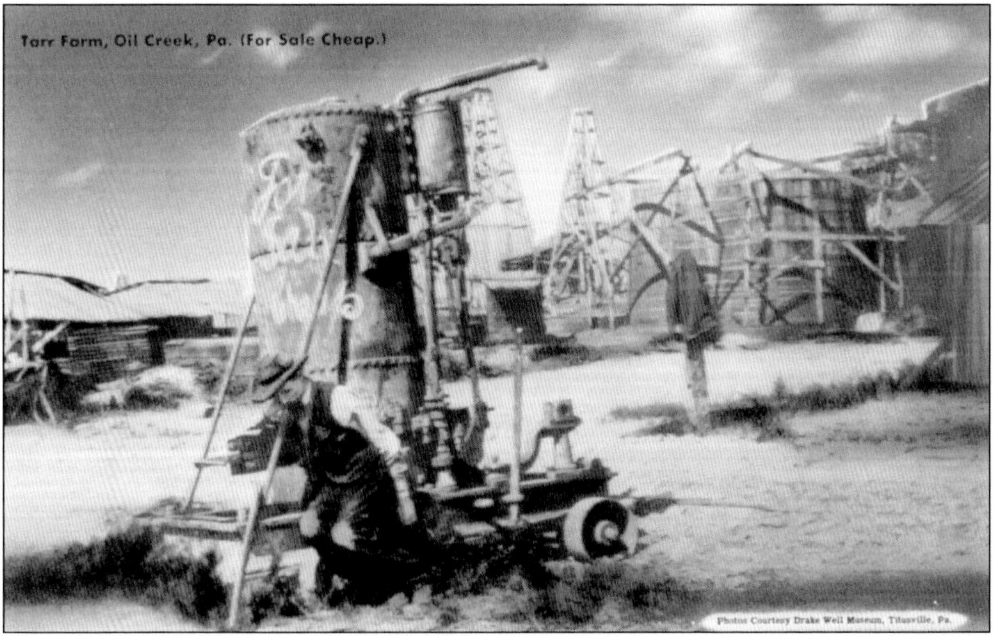

Oil fortunes began to dwindle in the Oil Creek and Pithole fields by the late 1860s and early 1870s, spurring exploration and speculation by wildcatters elsewhere in the region. The Tarr Farm on Oil Creek leaped into production in October 1861 when the Phillips Well started at 4,000 barrels daily. But by March 1865, oil production had ceased and images like the one above from the Tarr Farm—drilling apparatus "for sale cheap"—became common.

Two

The Oil Industry Expands Its Boundaries

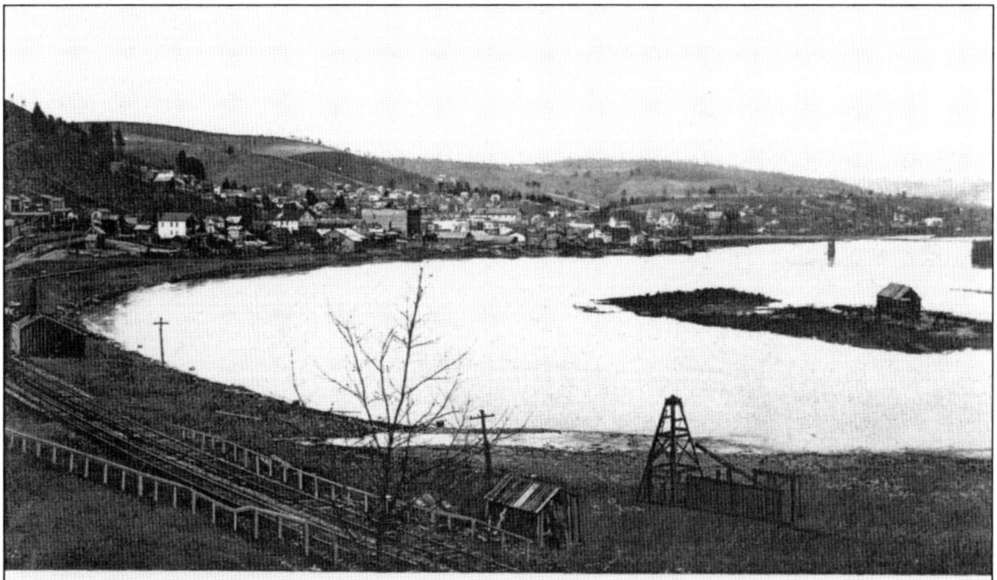

Greeting from Tidionte, Pa. One of the Oldest Oil Towns in the State

Tidioute was an established lumber town on the Allegheny River at the time of the early oil boom. The first oil well near Tidioute was sunk on Gordon Run by J. L. Grandin only two days after Edwin L. Drake's success at Oil Creek. The well came up dry. The King and Ferris Well, drilled below Gordon Run in July 1860, was a successful venture but only produced 10 barrels of oil daily.

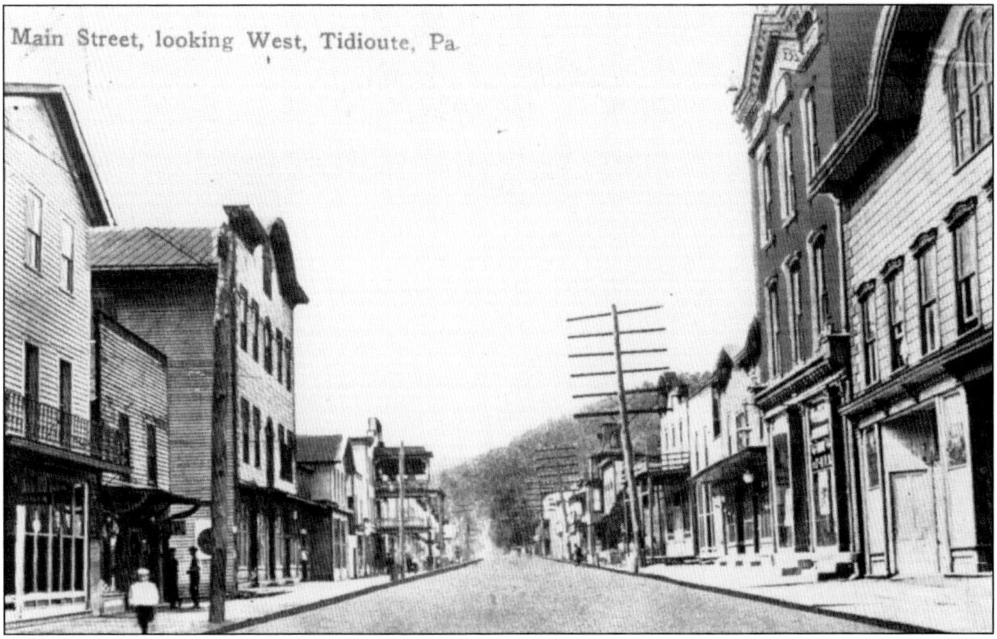

By the end of July 1860, over 60 wells had been drilled around Gordon Run. In August 1860, the first flowing well was drilled across the river from Tidioute on the Cohell farm. Rev. C. L. Hemquembourg and associates from Warren leased the land, drilled for oil with a spring pole, and fitted the well with wooden pipe and a farmhouse pitcher pump. The well would eventually produce 100 to 300 barrels of oil daily.

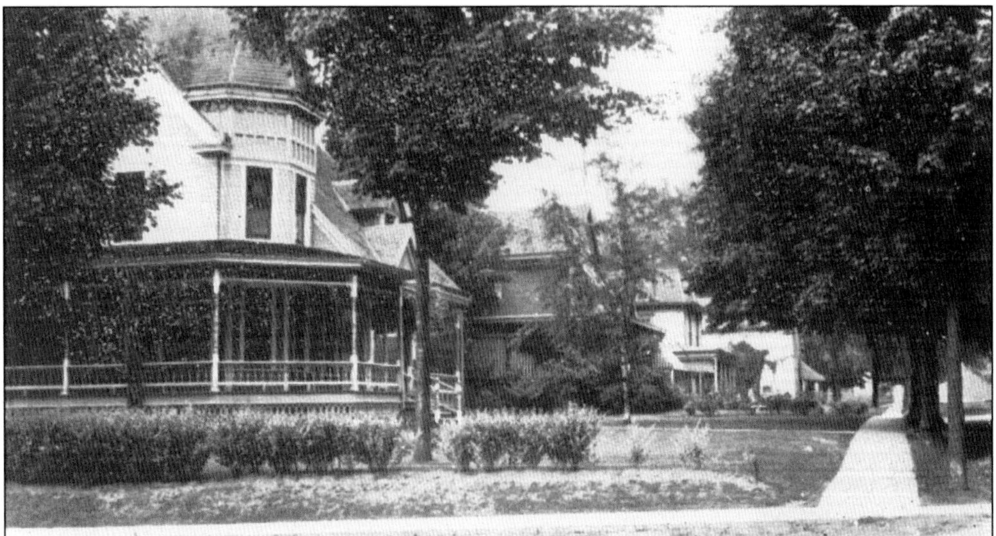

The Economites, a religious millennial group, owned 6,000 acres of land across the river from Tidioute. Drilling commenced in the summer of 1860. Between 1860 and 1873, 75 wells were drilled, the best producing over 600 barrels of oil daily. The end result of the oil boom was the rapid growth of the town. But like Pleasantville, Tidioute retained its civility unlike boomtowns, exemplified by this image of a quaint Victorian Main Street.

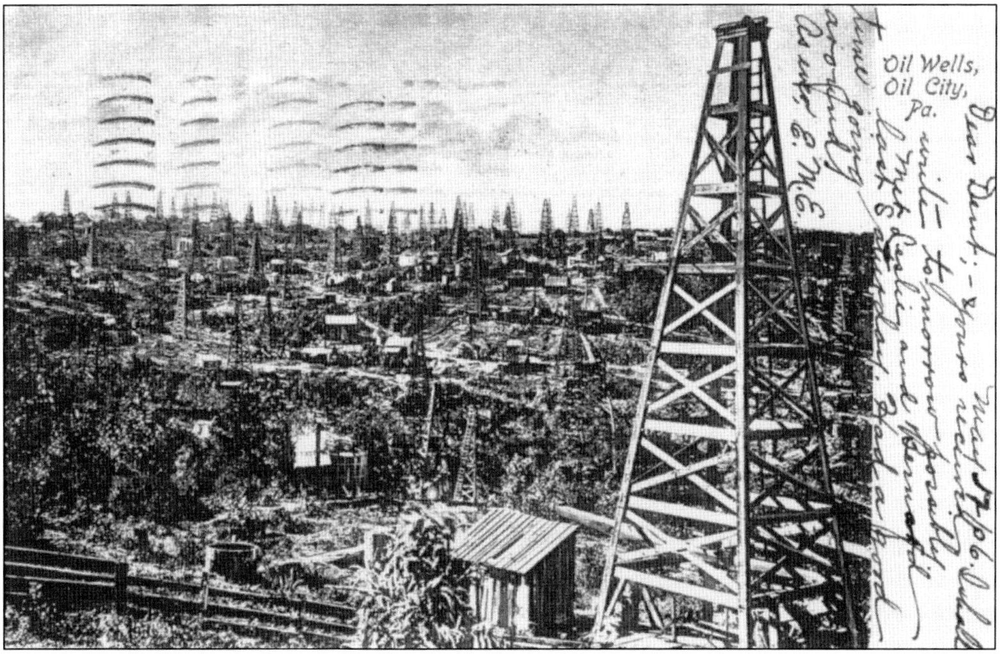

Triumph Hill, the highest point near Tidioute, was the site of considerable oil frenzy. The Triumph Hill development lasted about five years and at its peak had 3,000 inhabitants and the highest density of derricks anywhere in the oil region. In the summer of 1867, average oil production at Triumph Hill was 2,000 barrels daily. The discovery of oil near Pleasantville, well flooding, and high well density led to the quick demise of Triumph Hill.

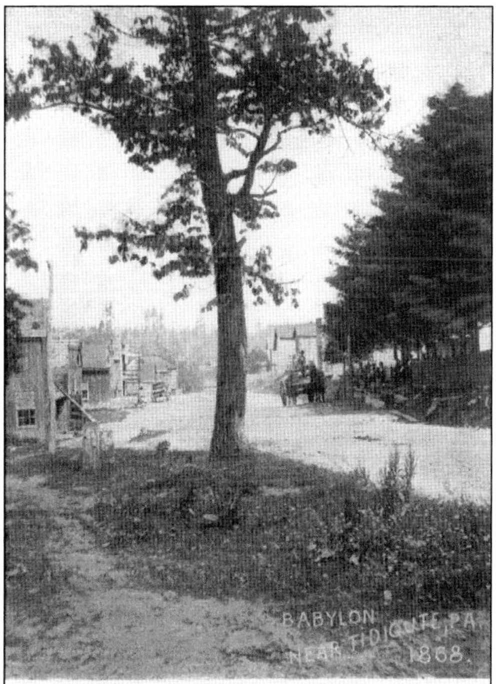

Babylon near Triumph Hill was not an oil town: six of its eight businesses were "sporting houses" or brothels. The infamous Ben Hogan, of Pithole fame, established a dance house and "gymnasium" at Babylon in the summer of 1866 and did a thriving business. This image of Babylon from 1868 shows the Warren-Franklin turnpike passing through the center of town and the ramshackle buildings flanking it.

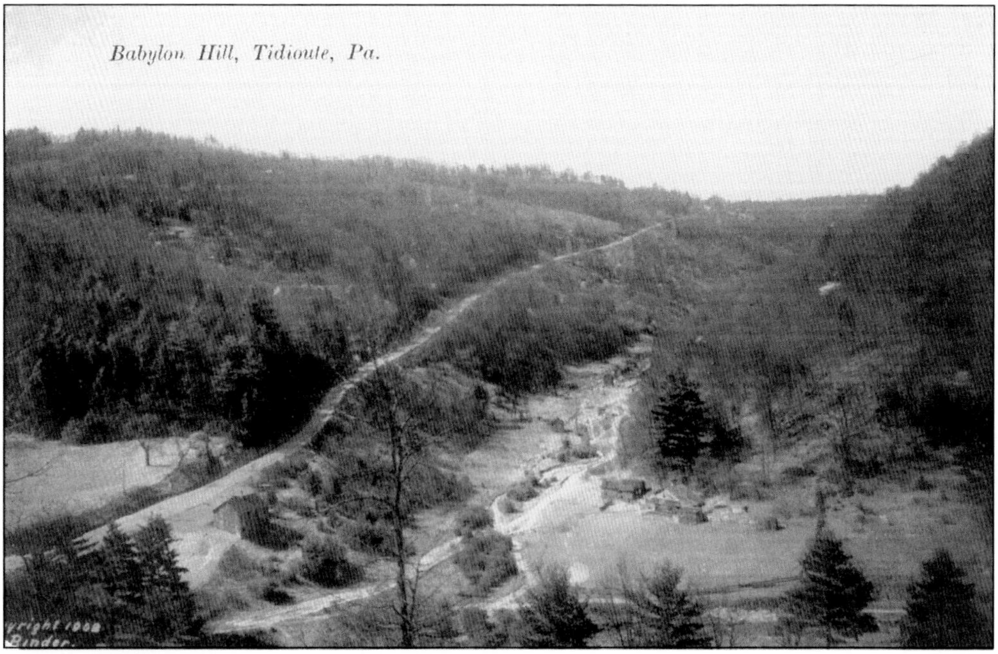

Babylon Hill is where the notorious red-light district of Babylon sprang up. By the early 1900s, Babylon was just a memory and a place name on maps. This view of Babylon Hill from the early 1900s shows a landscape recovering from the oil boom nearly 40 years earlier.

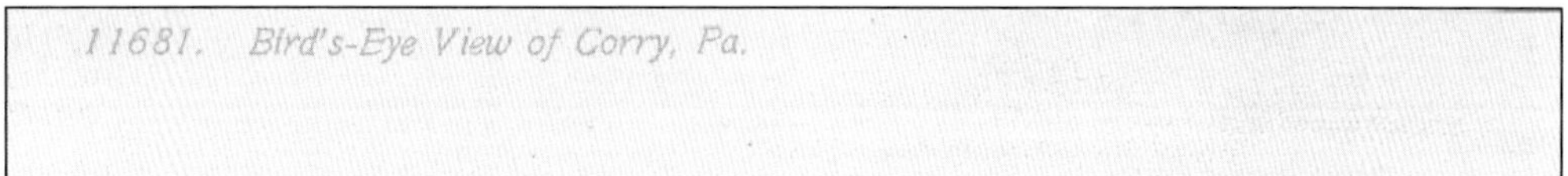

Located on Brokenstraw Creek in eastern Erie County, Corry was not in the center of the oil-producing region. In 1859, it was a railroad crossroads with a few houses. In 1862, the Oil Creek Railroad was completed from Corry to Titusville, and Corry became an important portal to the oil region. Known as the "city of stumps" due to its rapid growth, Corry had a population of 3,000 in 1865.

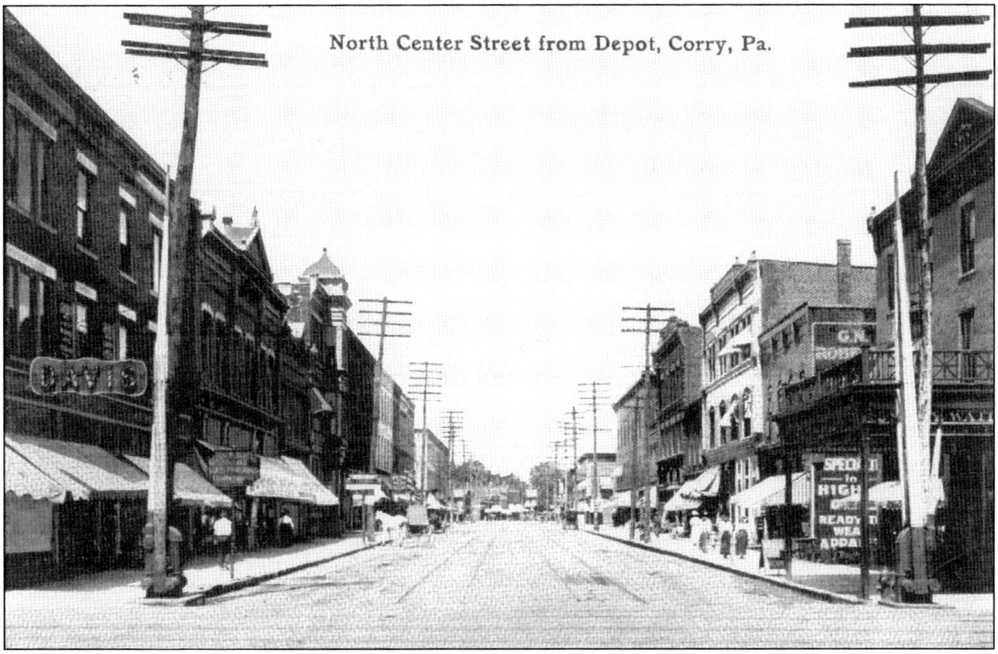

Besides being a railroad hub, Corry was also known as an oil-refining town. In 1861, the Downer Oil Works was established at Corry. When the refinery came online, it employed over 150 workers and had a capacity for refining a million gallons of oil yearly.

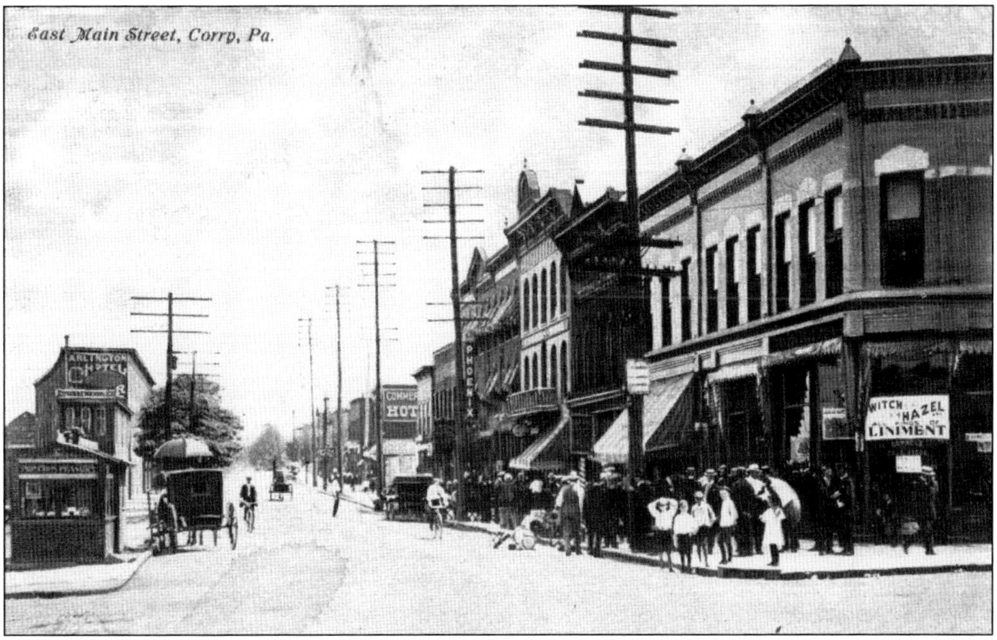

Corry's prominence as a railroad town led to the development of a diverse economy. In the early 1900s, Corry had over 30 industries. Among the most important were the Climax Manufacturing Company's locomotive works, the U.S. Radiator and Boiler Company, the Ajax Iron Works, the Corry Condensed Milk Company, the Corry Chair Company, the U.S. Chair Company, and the McInness Steel Company. This card, dated 1911, shows a view of East Main Street in Corry.

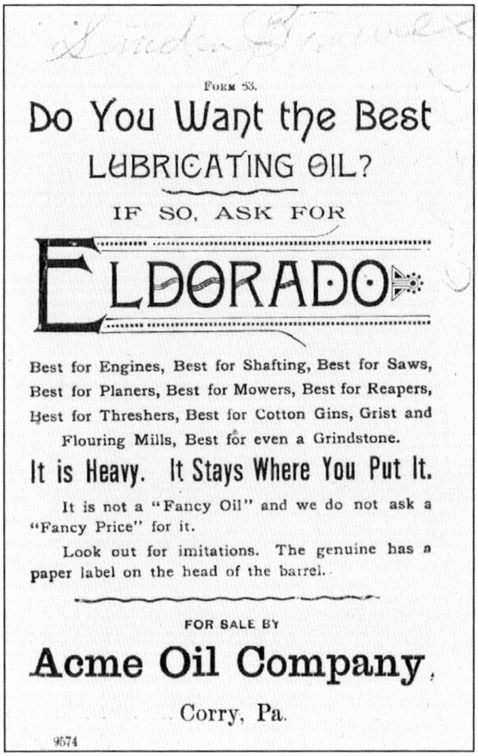

The Acme Oil Company, a Standard Oil affiliate, opened in Titusville in 1875 by purchasing four refineries that became Acme Nos. 1 to 3. On June 11, 1880, lightning struck a nearby pipeline, starting a fire that lasted three days, exploding storage tanks, and engulfing Acme No. 1. Acme Oil did not rebuild in Titusville but built a new plant in Olean, New York. Acme's Eldorado product, shown on this advertising card, was probably heavily used in industrial Corry.

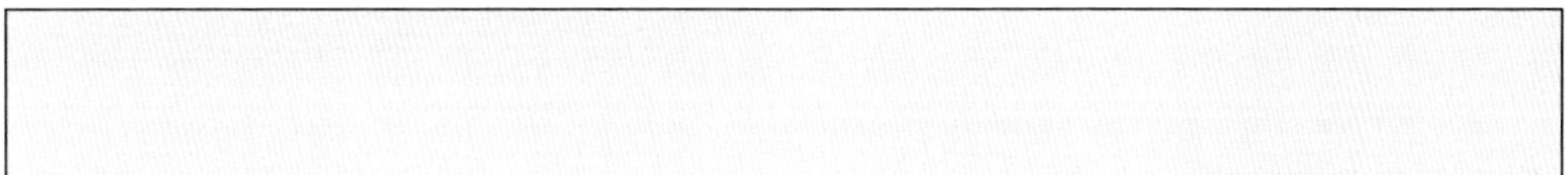

The rise of the Bradford oil field in the 1870s and the competition and increasing expense to prospect there led many oilmen to search elsewhere. In 1875, oil was struck at Glade near Warren, close to the confluence of Conewango Creek and the Allegheny River. An oil field of approximately 6,000 acres was identified in and around Glade and Warren that produced oil of high quality.

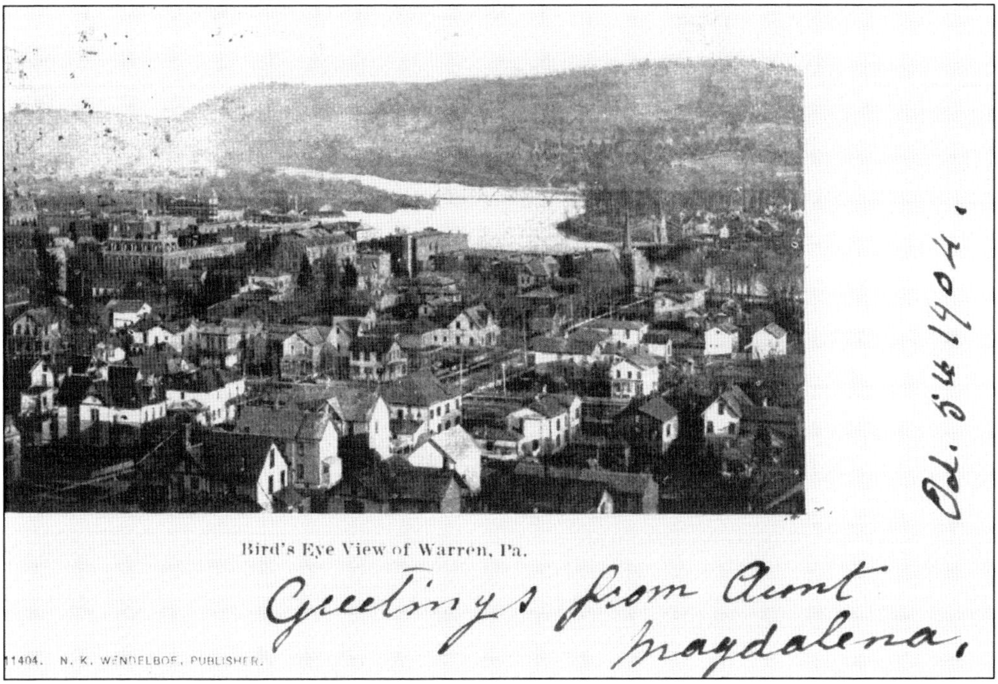

The city of Warren, located at the junction of the Allegheny River and Conewango Creek, was surveyed in 1797. In the early 1900s, Warren had a population of 11,000 and was considered a railroad center and an oil production hub with five refineries located in and around the town. The bird's-eye view of Warren was postmarked in 1904.

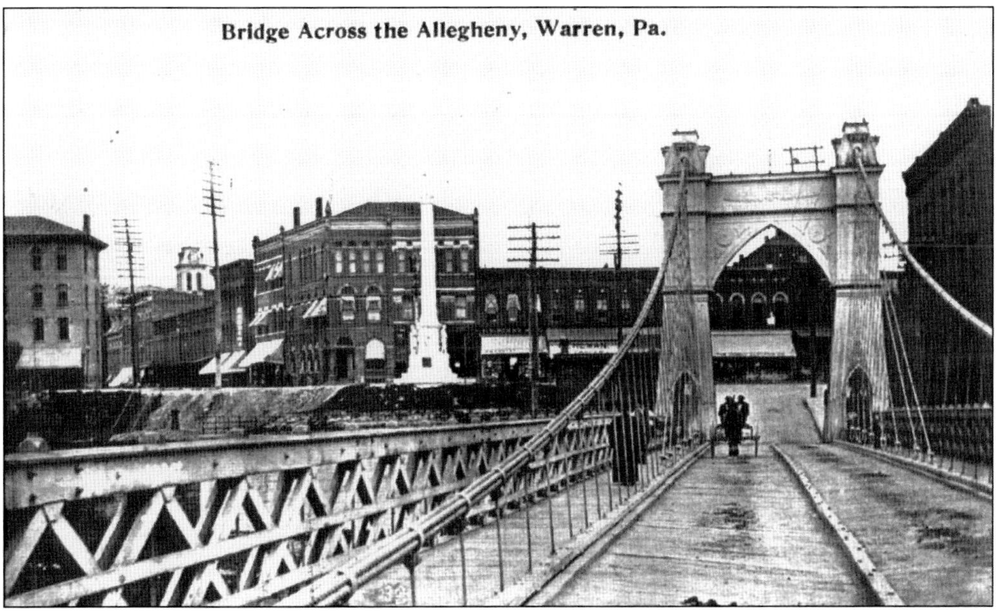

This postcard, titled "Bridge Across the Allegheny, Warren, Pa," shows the suspension bridge that formerly spanned the Allegheny River. It was constructed in 1871 and taken down in 1918 with the completion of the new concrete Hickory Street Bridge. The Soldiers and Sailors Monument, dedicated in 1909, can be seen to the left of the image.

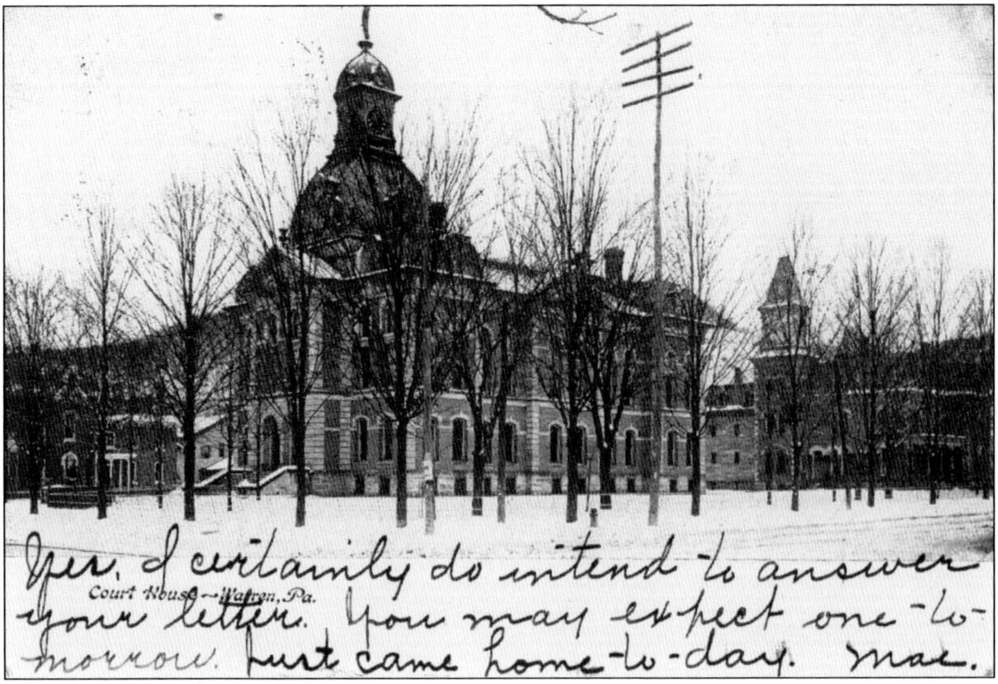

The Warren County Courthouse, seen in this postcard postmarked 1907, was designed by architect M. E. Beebe and built in 1877. It replaced the previous structure built in 1825. The Warren County Courthouse is one of the few county courthouses in Pennsylvania that is located in a residential area and not a business area.

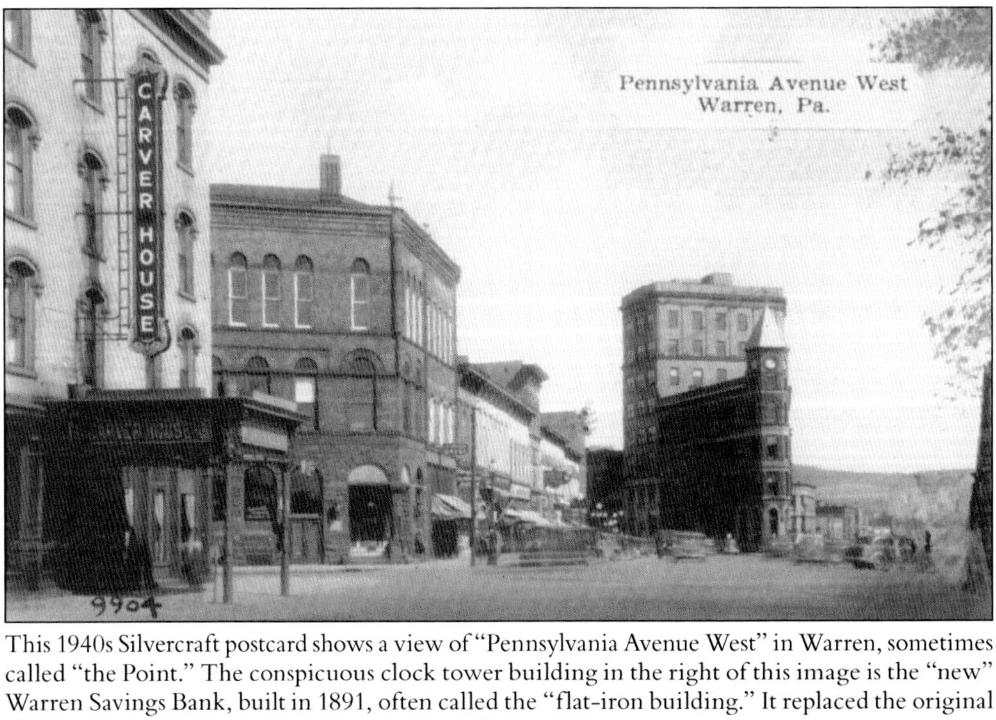

This 1940s Silvercraft postcard shows a view of "Pennsylvania Avenue West" in Warren, sometimes called "the Point." The conspicuous clock tower building in the right of this image is the "new" Warren Savings Bank, built in 1891, often called the "flat-iron building." It replaced the original Warren Saving Bank that was destroyed in a fire in 1889.

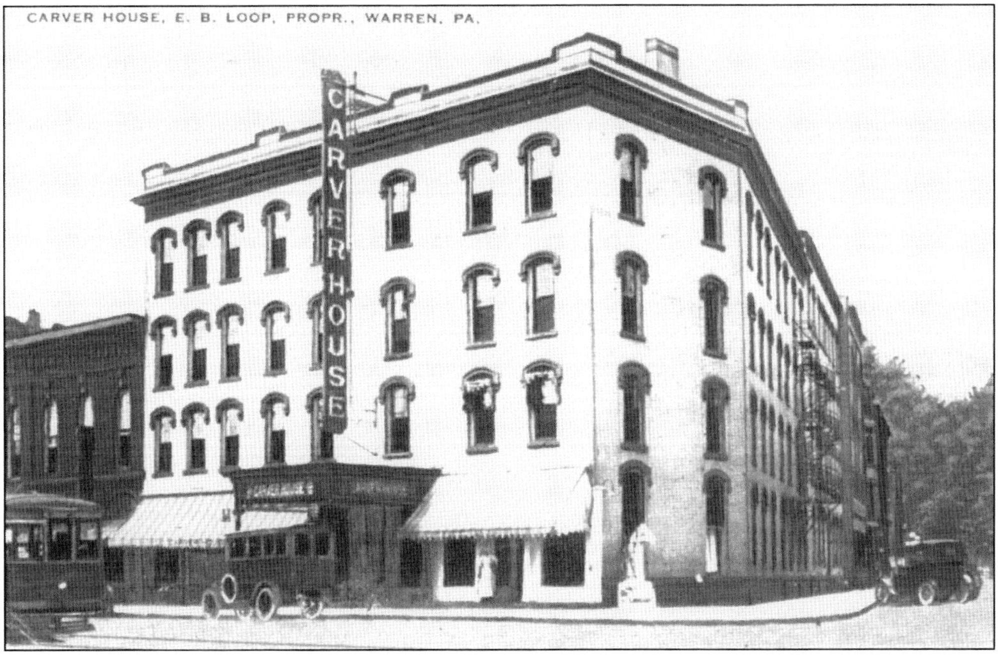

The Carver House Hotel was built in 1849 at the interaction of Pennsylvania Avenue and Hickory Street. It was destroyed in a fire in 1956. E. B. Loop is listed as proprietor of the Carver House in this 1920-era postcard.

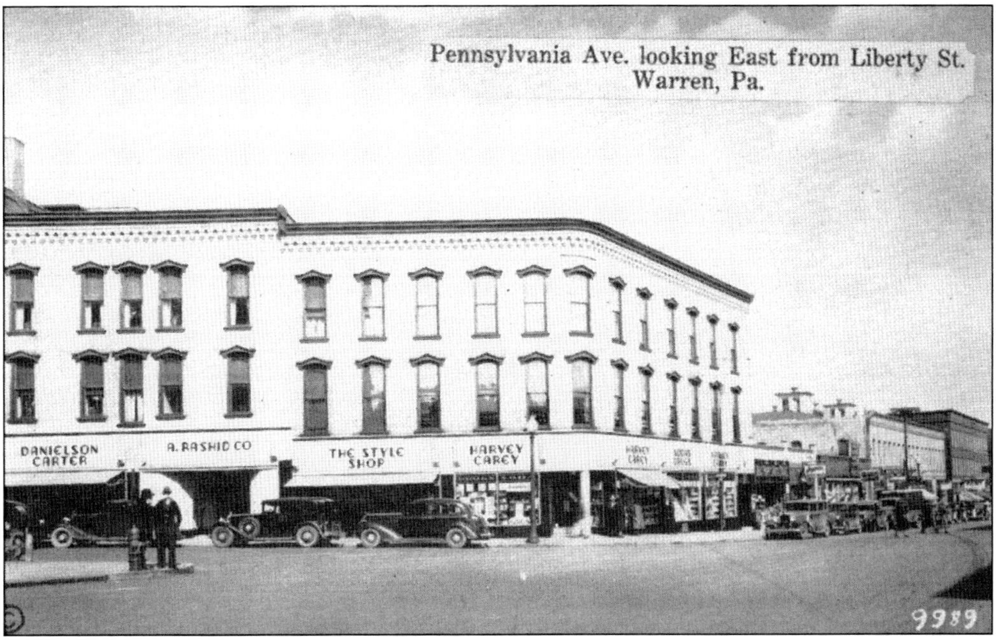

This view of "Pennsylvania Avenue Looking East from Liberty St." shows a busy downtown section of Warren in the 1940s. From left to right, shops in this image include Danielson Carter, A. Rashid Company, the Style Shop, and Harvey Carey.

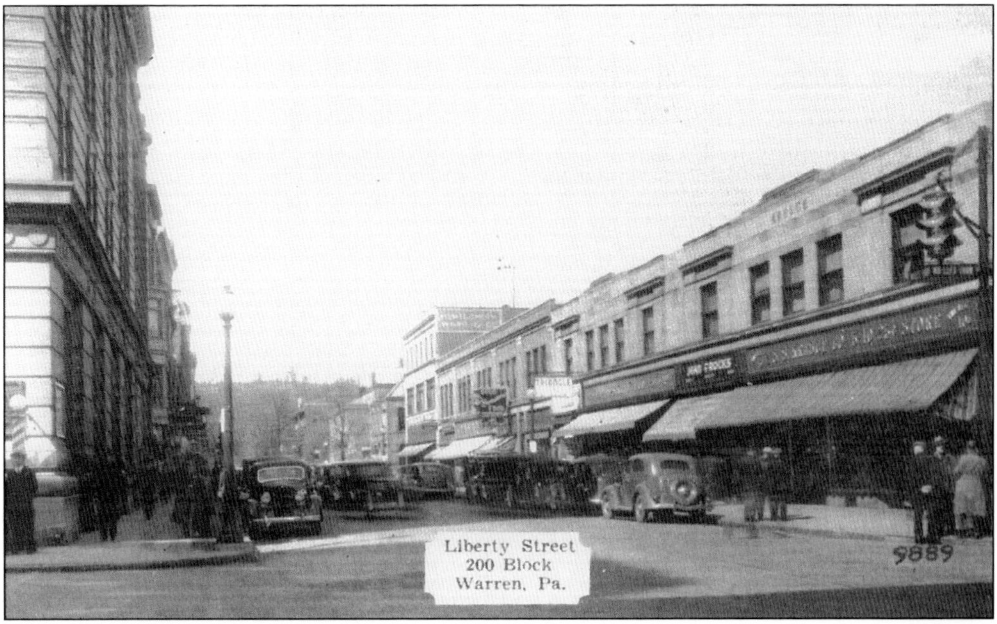

The busy 200 block of Liberty Street in Warren is showcased by this Silvercraft postcard from the 1940s. Numerous shops front the street, including a five-and-dime in the right of the image. Parking seems to be at a premium.

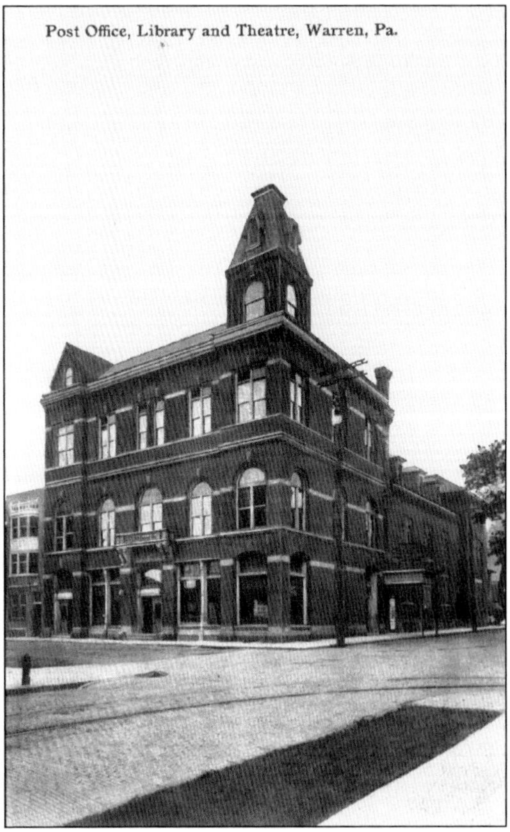

Warren resident Thomas Struthers was well known as a lawyer, real estate magnate, and legislator. In 1882, Struthers offered to construct a building that would be a cultural center for Warren and its residents. The Struthers Library Building opened in 1884. The building housed a library, stores, a post office, and the "Library Hall," an opera house and ballroom with a seating capacity of over 900. This postcard of the Struthers Library Building was postmarked in 1911.

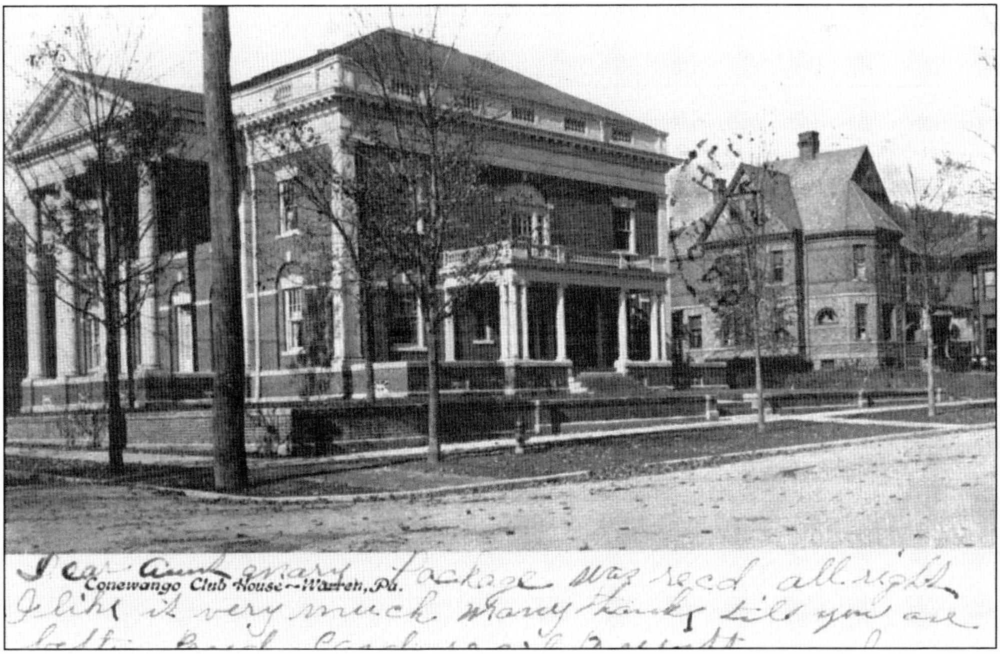

The Conewango Club is a historic building located on Market Street in Warren. It has served the community as a place for social gatherings and receptions. This view of the Conewango Club was postmarked in 1907.

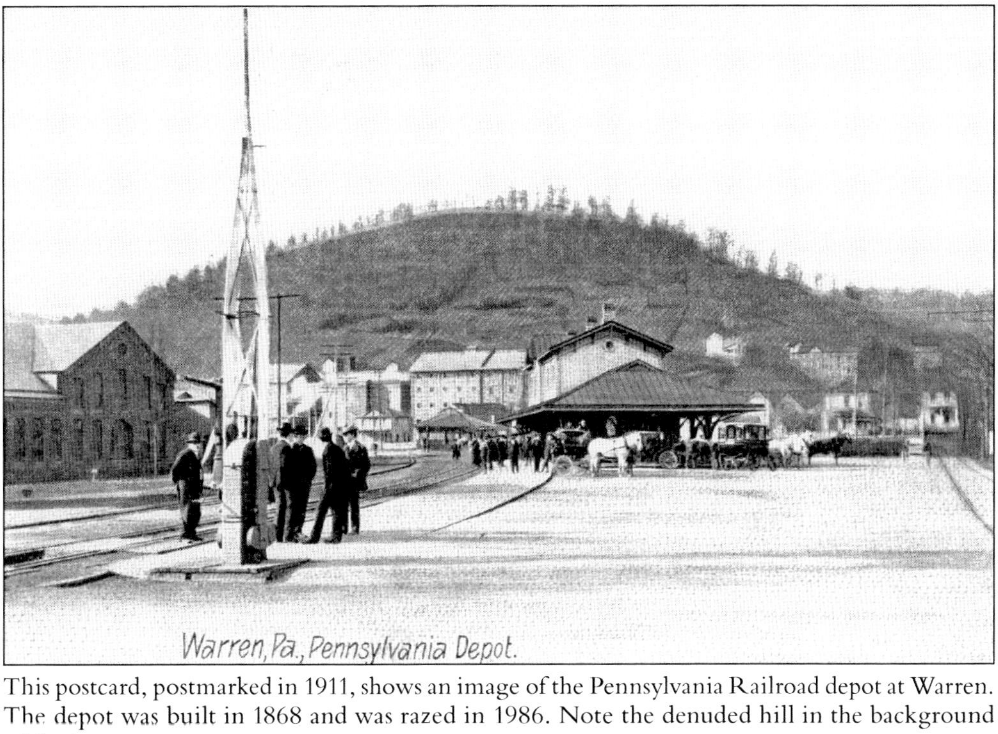

This postcard, postmarked in 1911, shows an image of the Pennsylvania Railroad depot at Warren. The depot was built in 1868 and was razed in 1986. Note the denuded hill in the background of the image.

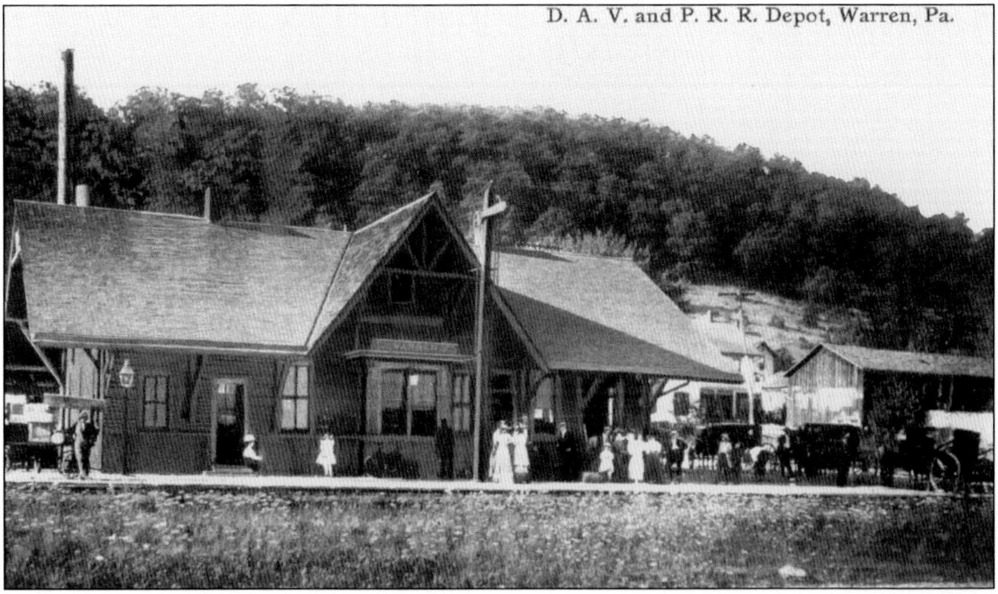

The Dunkirk, Allegheny Valley, and Pittsburgh Railroad (DAV&PRR) was formed by the merger of the Dunkirk, Warren and Pittsburgh Railroad Company with the Warren and Venango Railroad Company in 1872. The DAV&PRR line ran from Dunkirk, New York, to Titusville. This view of the DAV&PRR depot in Warren was postmarked in 1912. It was located at the corner of Laurel Street and Fourth Avenue until passenger service was discontinued in 1937, and the depot was razed.

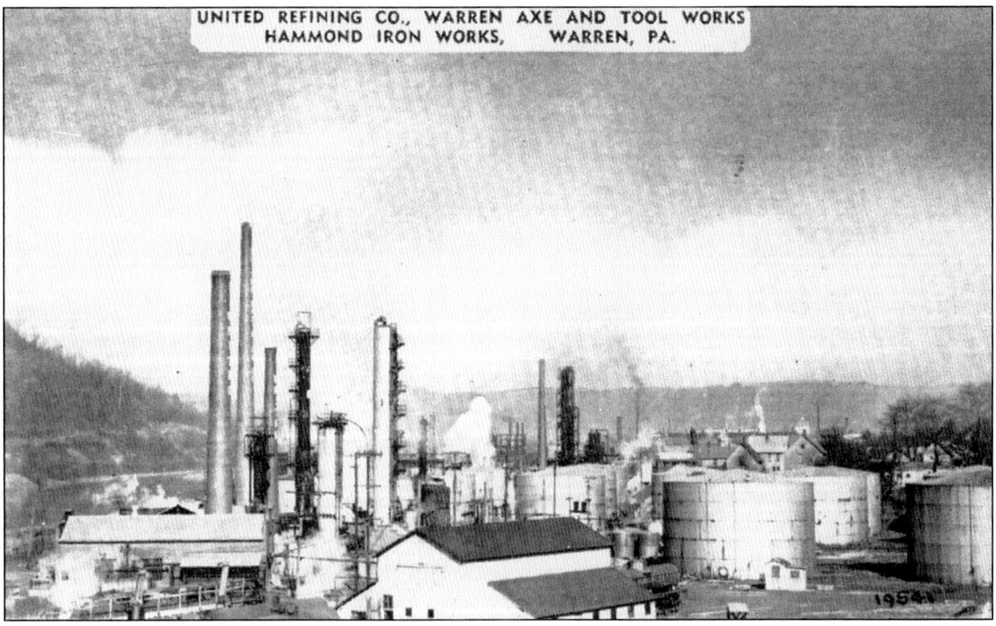

The United Refining Company's plant in Warren was founded by Harry Logan Sr. in 1902. Located on the Allegheny River near Glade, the plant is still in operation and currently processes 70,000 barrels of oil daily. This Silvercraft postcard from the 1940s shows the United Refining Company's plants as well as the Warren Axe and Tool Works and the Hammond Iron Works, which made oil tank cars and storage tanks.

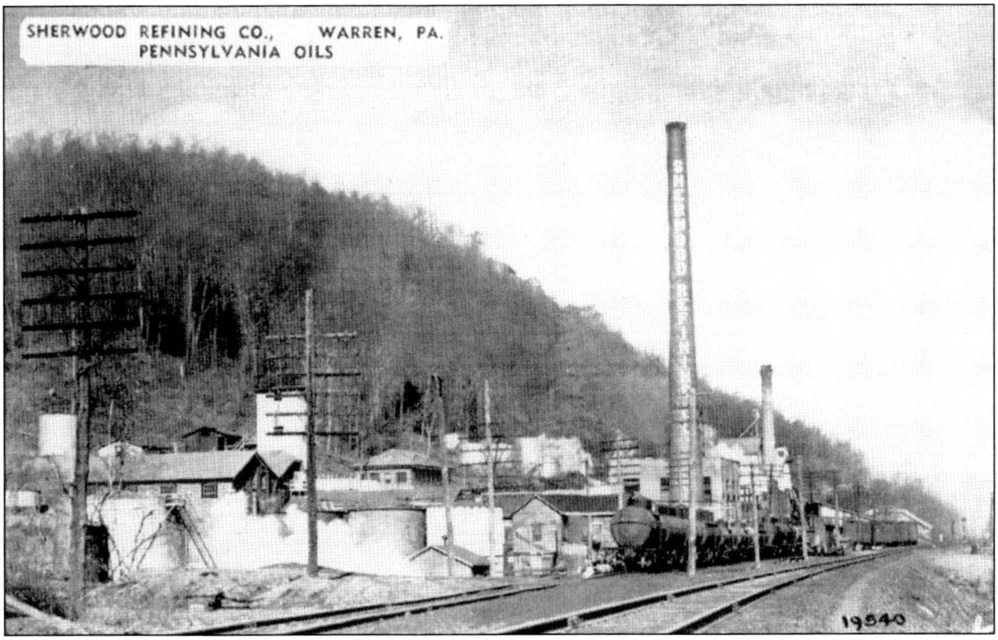

The Sherwood Refining Company was located on the south bank of the Allegheny River at Warren. The caption on the back of this 1940s Silvercraft postcard notes that "White mineral oils, petrolatum and their derivatives are the principal products of this modern and growing refinery . . . It provides regular employment for about 60 persons. Main offices are in Englewood, New Jersey and products are sold throughout the world".

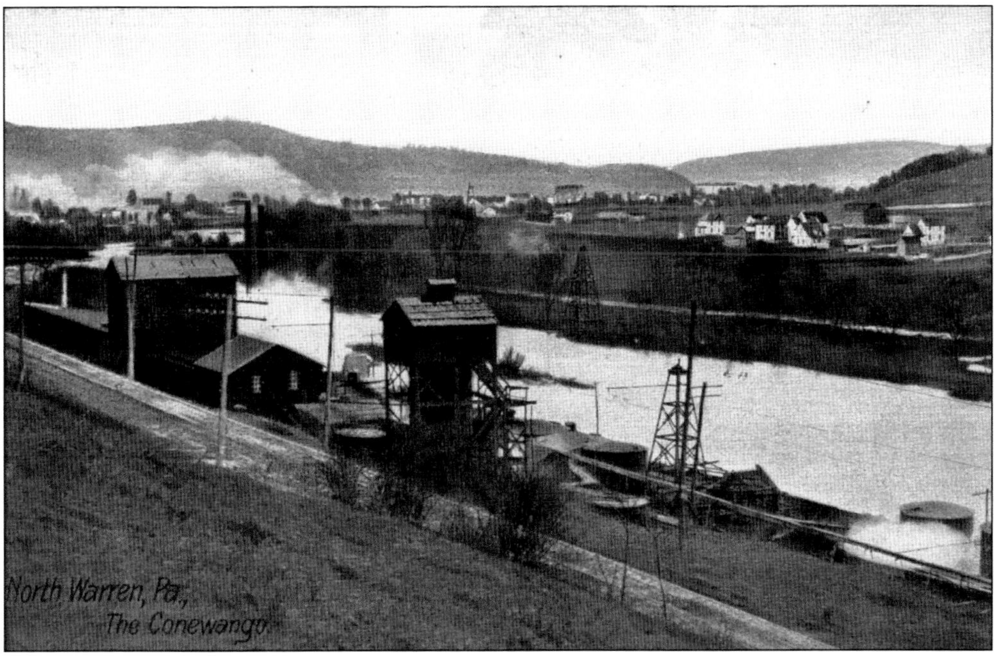

North Warren is located along Conewango Creek. This view of North Warren, postmarked 1912, shows oil storage facilities and a few derricks on the banks of the creek. Oil development along Conewango Creek has been generally light.

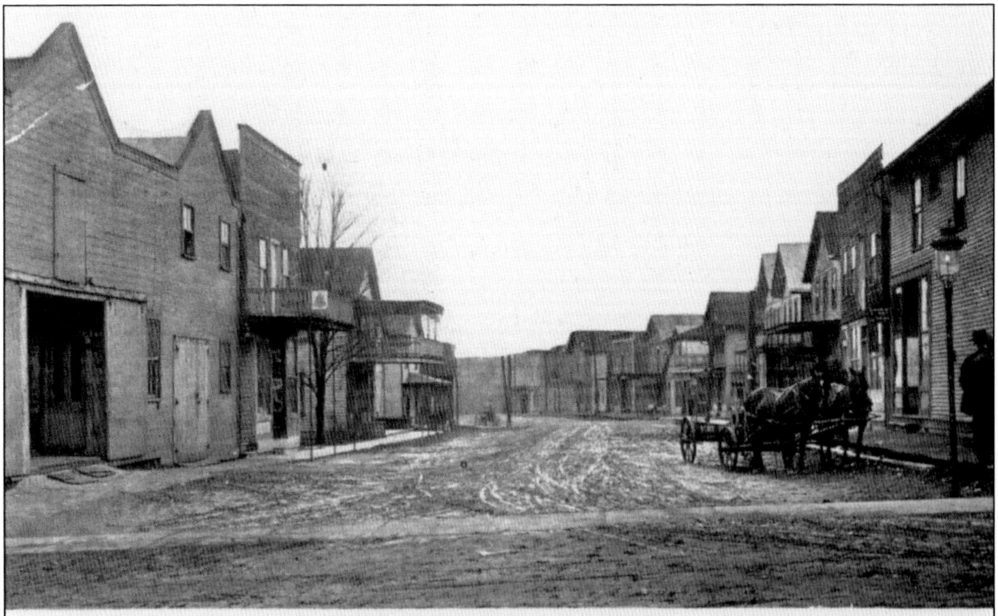
RAILROAD STREET, Clarendon, Pa.

The Clarendon Oil Field is located south of Warren and encompassed 24,000 acres along the West Branch of Tionesta Creek. Drilling in the Clarendon Oil Field began in 1878, and the first well drilled in Clarendon was the Egan Well in 1880. The Clarendon Oil Field was known as a steady producer of high-quality lubricating oil. This postcard, postmarked 1907, shows an image of downtown Clarendon along Railroad Street.

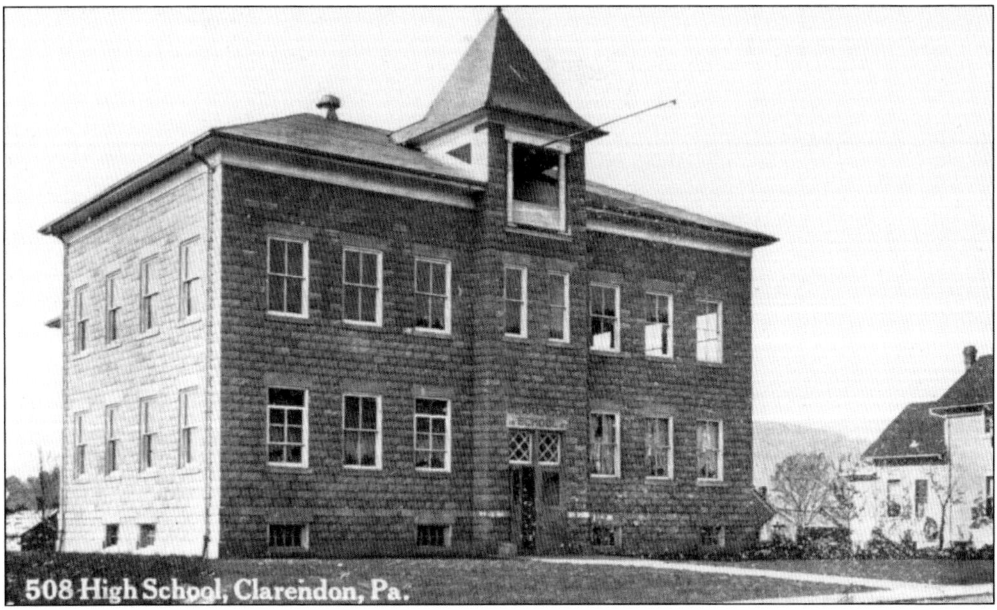
508 High School, Clarendon, Pa.

In the early 1900s, Clarendon had a population of about 1,100. Oil and leather tanning were its chief industries. There were three oil refineries in the town, including the Levi Smith Refinery and the John Goal Refinery. The Levi Smith Refinery employed 15 to 20 persons and processed 8,000 barrels of crude oil monthly. This early-1900s image shows the high school at Clarendon, which now houses borough offices.

Three

The Bradford Oil Field

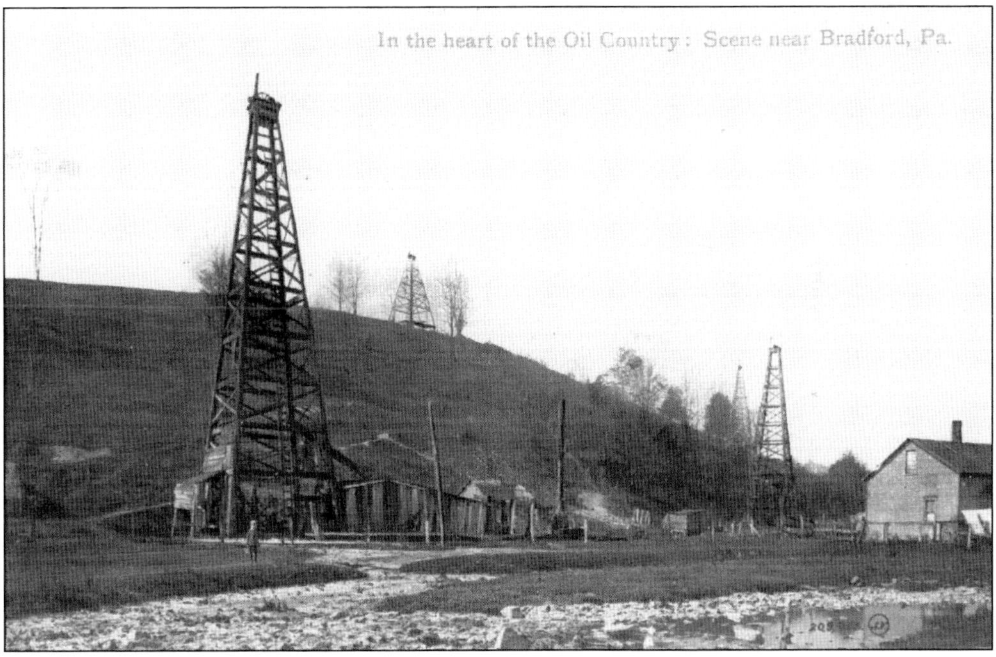

Bradford was a small lumbering village before oil changed its fortunes in the 1870s. The area around Bradford was settled in 1837 by Levitt Little, an agent for the United States Land Company, which controlled over 250,000 acres in the region. The small village of Littleton became Bradford in 1858. This postcard shows a quaint view of the oil country about Bradford in the early 1900s.

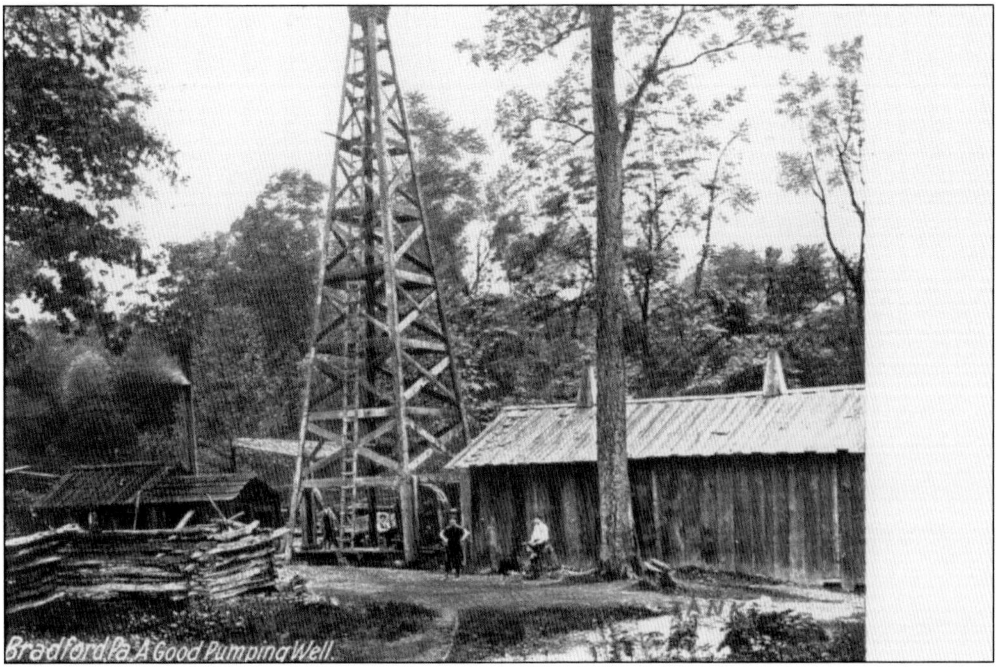

As early as 1868, wells of limited productivity were drilled in what would become the Bradford Oil Field. In 1875, the Crocker Well was opened, producing 300 barrels daily. Attention was quickly drawn to the Bradford area. The Bradford Oil Field would encompass 84,000 acres of land in Pennsylvania and New York. In 1881, the region would produce 90 percent of the nation's oil.

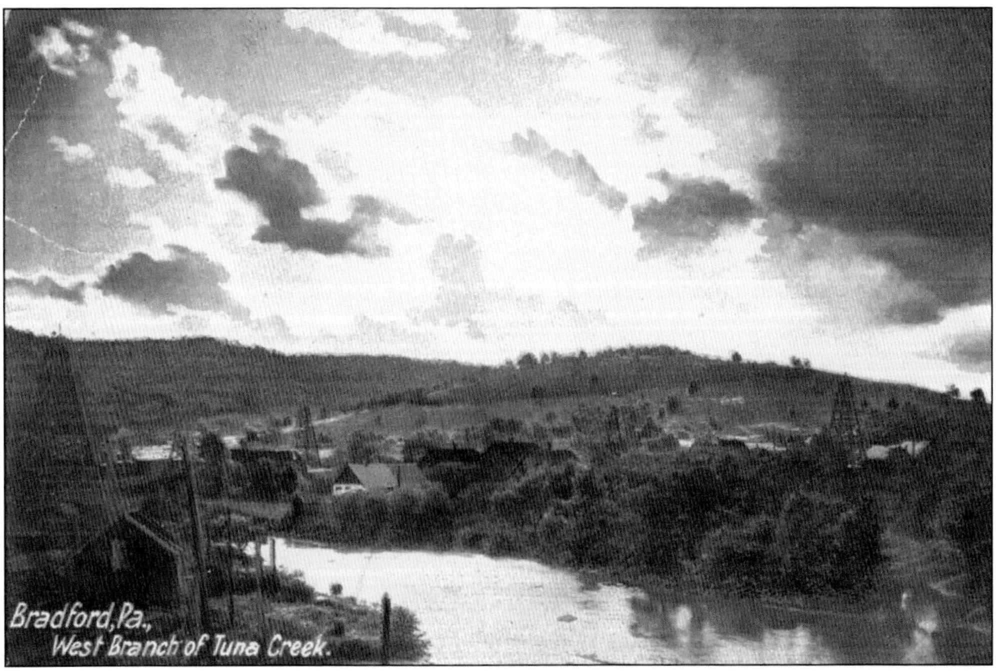

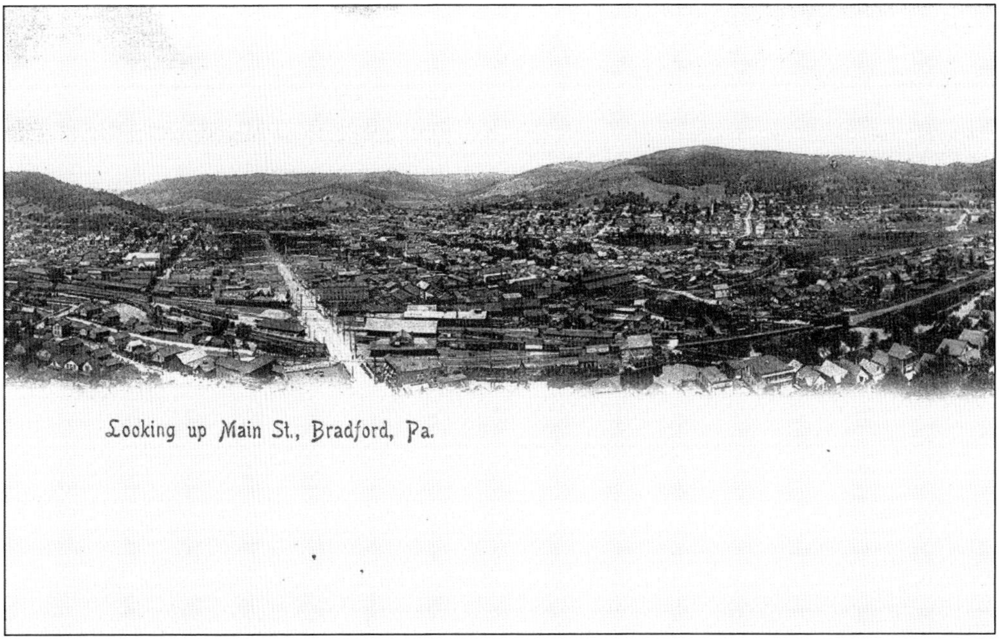

Bradford is located on Tunungwant Creek, a tributary of the Allegheny River in McKean County, just below the New York state line. In the early 1900s, Bradford had a population of over 17,000. At the time, Bradford hosted two oil refineries and five companies manufacturing oil well supplies.

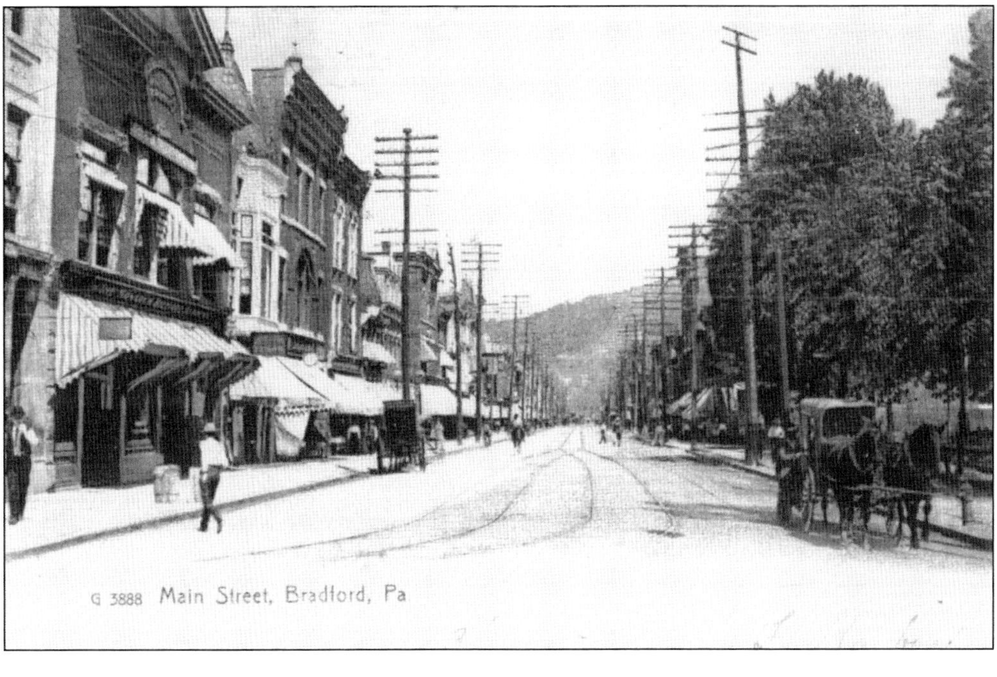

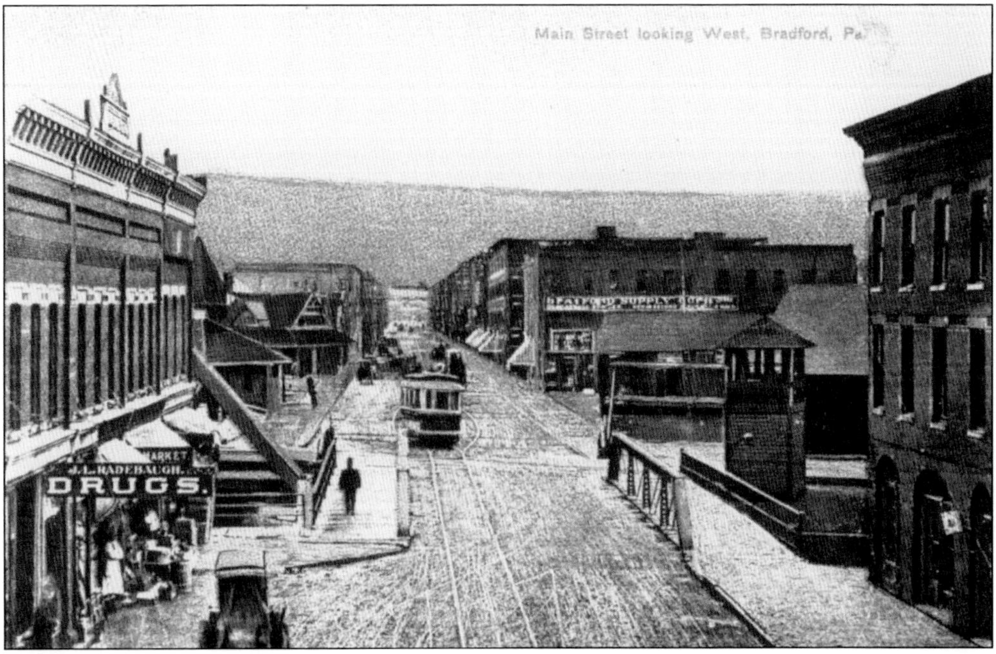

Bradford had a population of 500 to 600 residents in 1875. Shortly after the oil boom hit, J. J. McLaurin, an early oil historian, noted that Bradford had "Ten churches, schools, five banks, stores, hotels, three newspapers, street-cars, miles of residences, and fifteen-thousand of the liveliest people on earth." The postcard above shows an image of "Main Street Looking West," in the early 1900s while the postcard below, postmarked in 1907, shows the "Public Square" in Bradford.

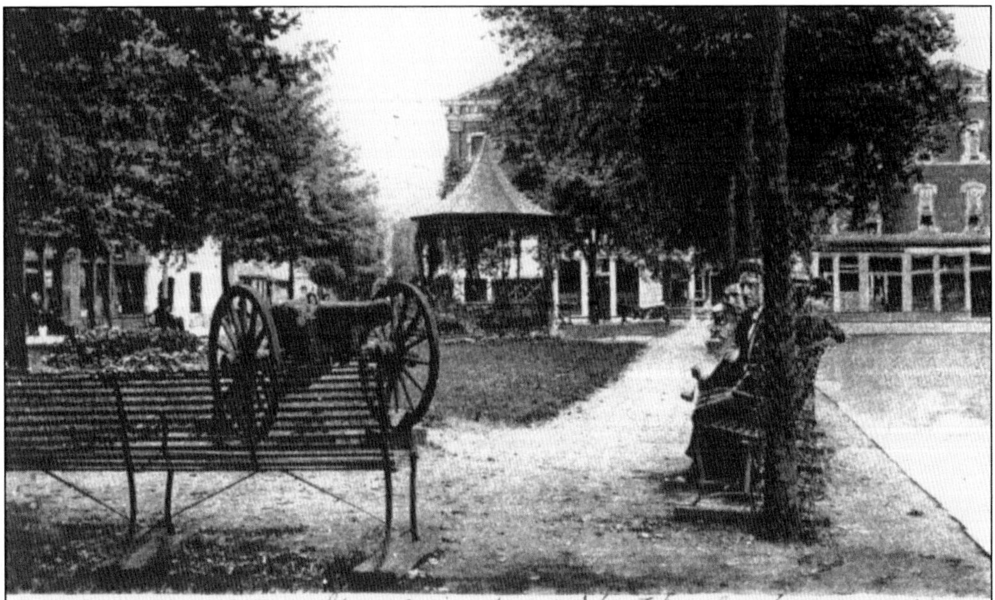

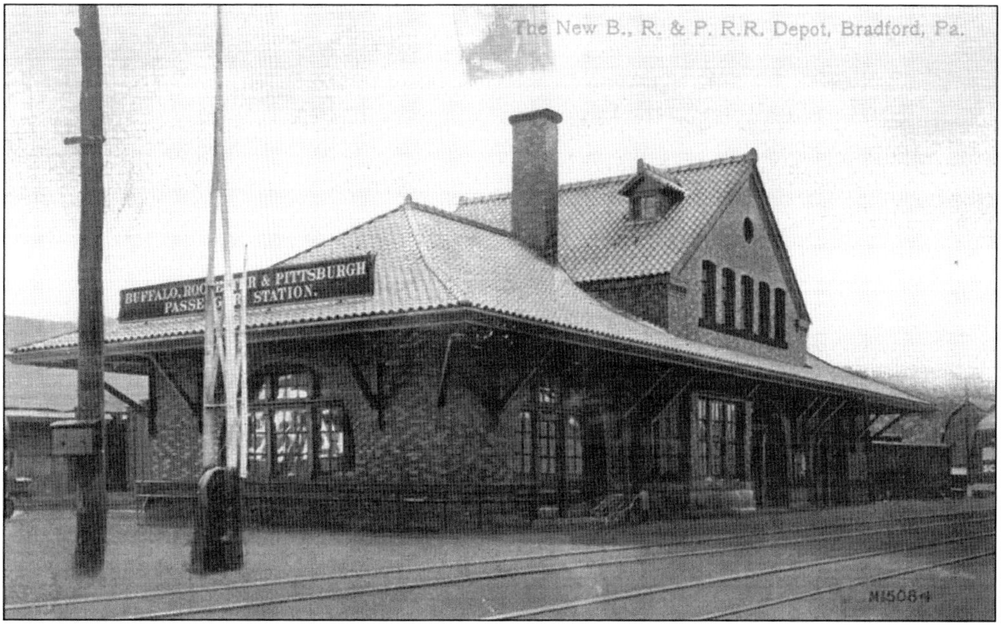

Like most of the towns that lived beyond the oil boom, Bradford had a diverse economy that included non-oil enterprises. Among the industries in Bradford in the early 1900s were silk mills, foundries, a glass factory, a terra-cotta works, and repair shops for the Erie Railroad and the Baltimore, Rochester and Pennsylvania Railroad. These two images from the early 1900s show the Baltimore, Rochester and Pennsylvania Railroad depot and the Erie Railroad depot.

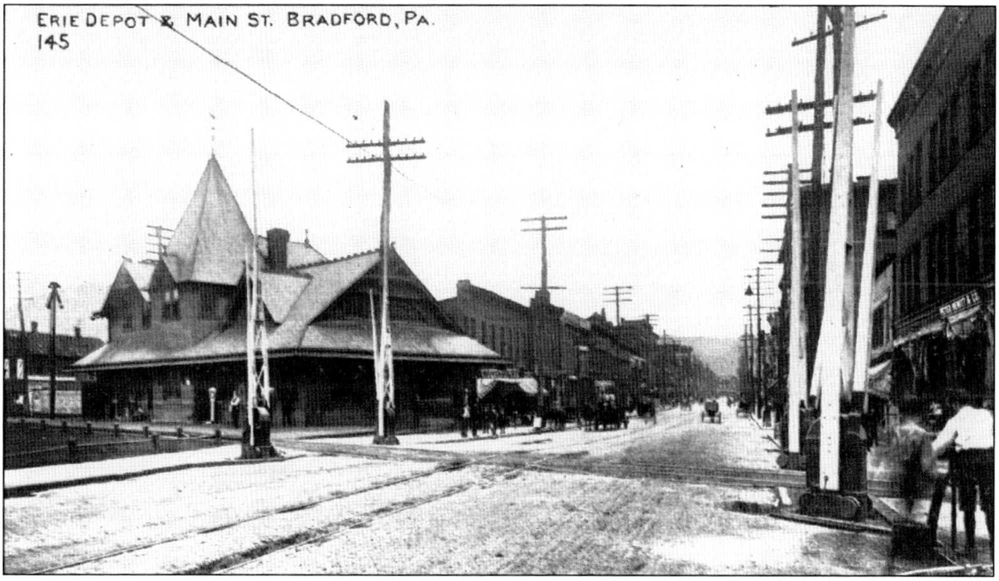

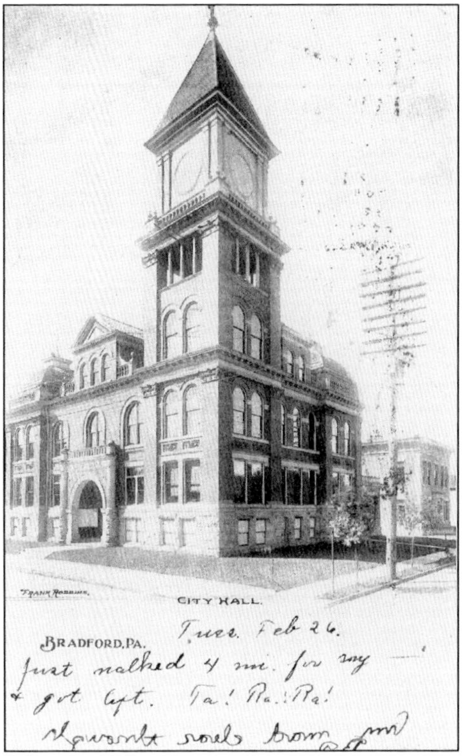

One of the venerable old buildings in Bradford is the old city hall located at the corner of Kennedy and Boylston Streets. Old city hall is listed on the National Register of Historic Places. The view of the building was postmarked in 1907.

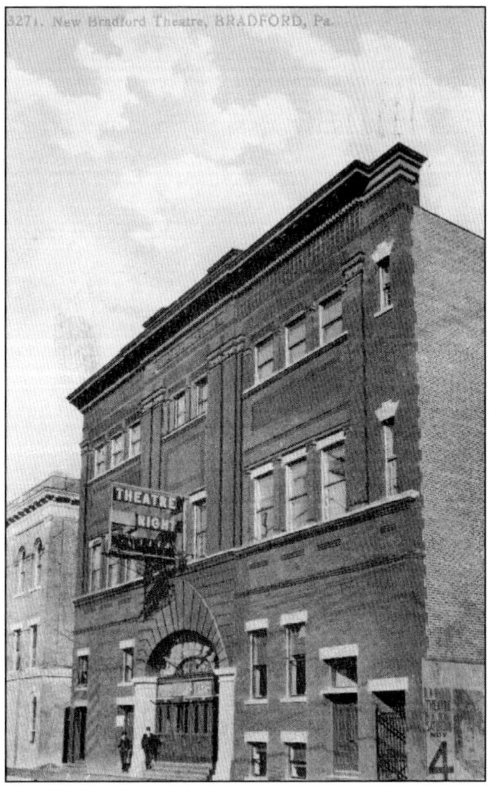

The Bradford Theatre was built in 1903, an endeavor promoted by local businessmen. *The Bradford Era* noted the "opening" in December 1903: "At 7 o'clock last evening the New Bradford Theater was formally opened for inspection by the public and for the auction sale of seats and boxes. Although an admission fee of 50¢ was charged, several hundred persons attended and the sale of seats was quite spirited for a time."

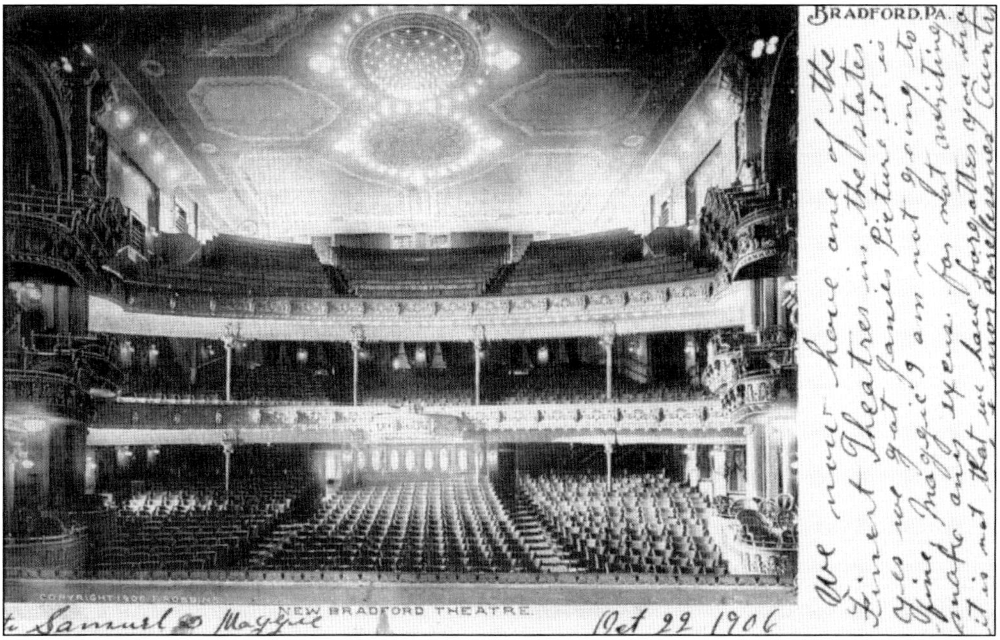

The Bradford Theatre was called "a temple of amusement" when it opened in 1903. It could seat 1,500. In 1922, the theater was sold to the Shea Theatre Company of Buffalo, New York. The building fell into disuse in the 1950s and was demolished in 1961. This postcard, dated 1906, shows the interior of the Bradford Theatre. The message on the back of this card reads, "we now have one of the finest theatres in the state."

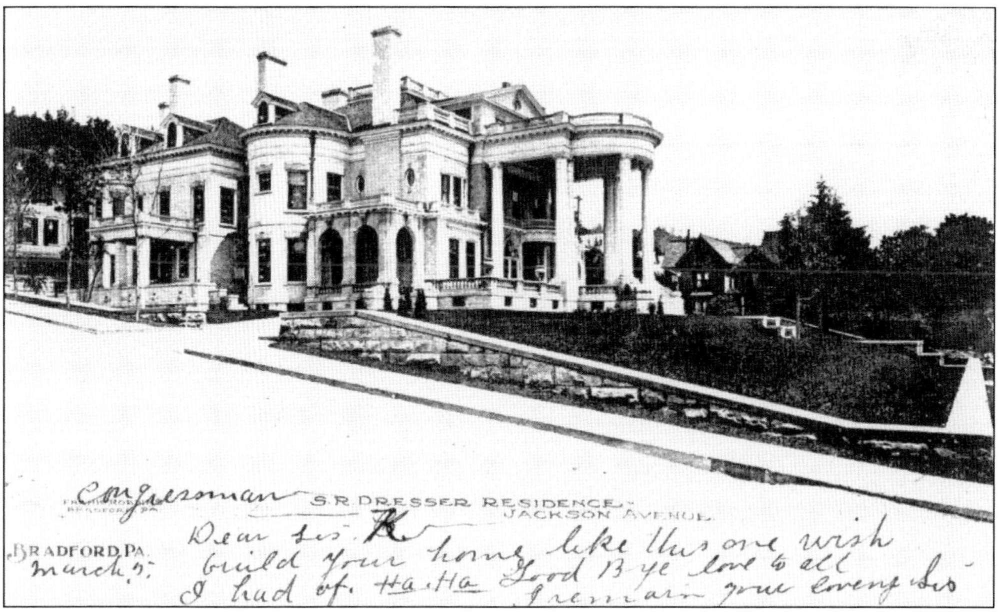

Solomon Dresser was an inventor, owner of S. R. Dresser and Company, and served in the Pennsylvania legislature. Dresser built this palatial home on Jackson Street in Bradford in 1903. Dresser thought his house "embodies all that is finest and best in a modern palatial residence . . . its quiet taste and dignified magnificence impresses every visitor to Bradford." The message on the back of this postcard reads, "build your home like this one, wish I had of."

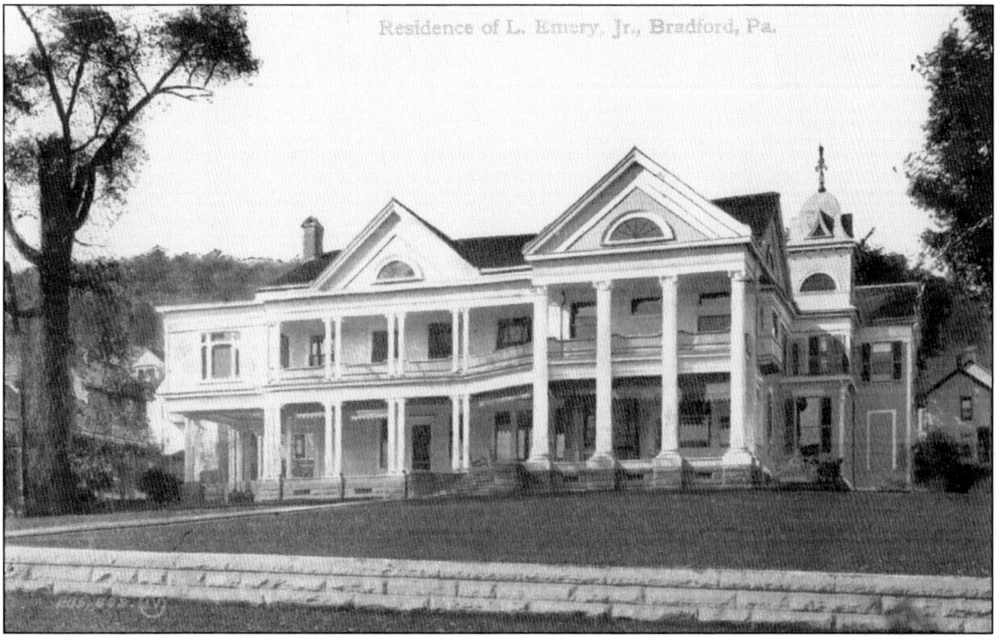

Lewis Emery Jr. entered the oil business at Pioneer on Oil Creek in the 1860s. He became bankrupt in 1873 and moved to the Bradford Oil Field, where he became a success and ultimately a millionaire. Emery was involved in numerous business ventures such as building refineries, huge hardware stores, and oil pipelines. Emery served in the state legislature and U.S. Senate. This postcard shows the Lewis Emery Jr. residence on Bradford's Congress Street in the early 1900s.

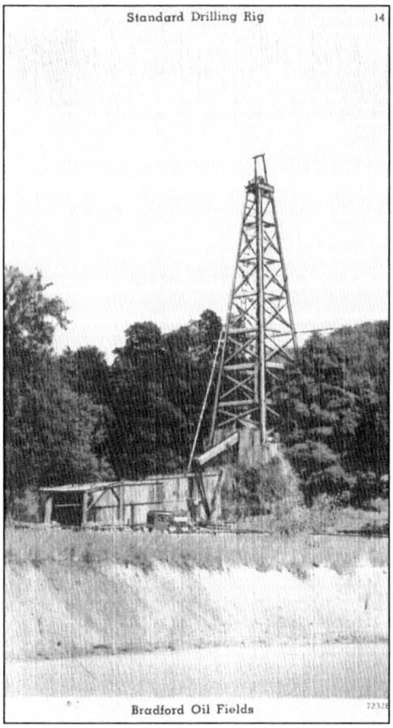

This linen-era postcard shows a standard drilling rig at the J. H. Gibson Estate near Bradford. The caption of the card reads, "The standard cable drilling rig consists of two principal parts: (1) The Derrick, a high pyramidal framed structure, erected directly over the site selected for the well; (2) A long narrow and comparatively low structure which houses the engine, the belt, a large band wheel and other mechanism provided for applying and controlling power."

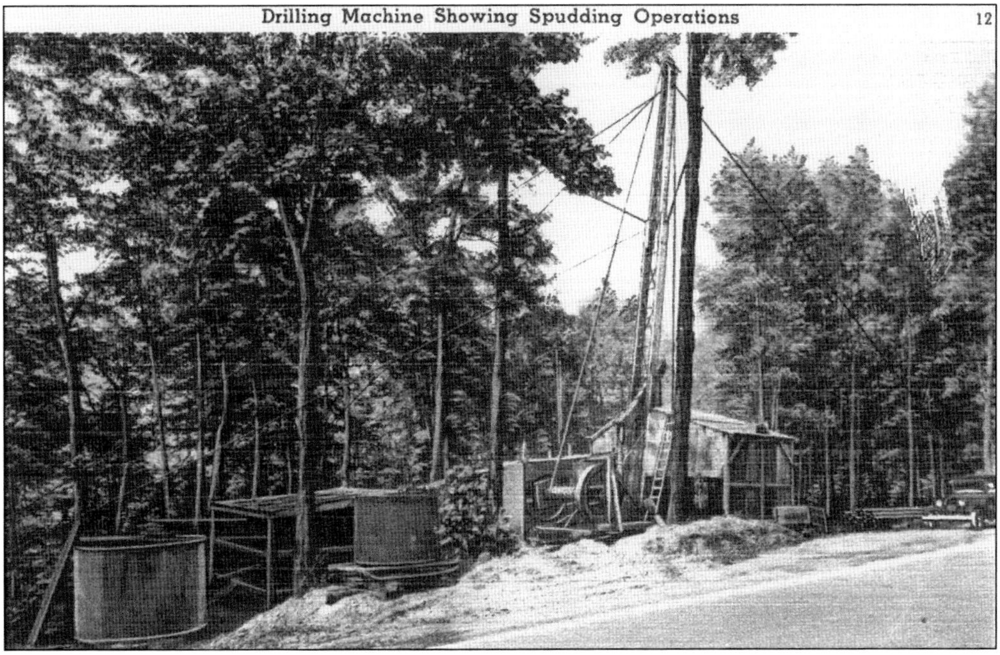

This linen-era postcard shows a drilling machine in the operation of "spudding" at the Wyborg Lease near Derrick City and Bradford. Spudding is the start of a new well. The caption of the card reads, "This type of rig employs the cable drilling system. A churning motion is imparted by means of the walking beam to the drilling tools, these in turn fracture or abrade the rocky formations which must be penetrated to reach the oil reservoir."

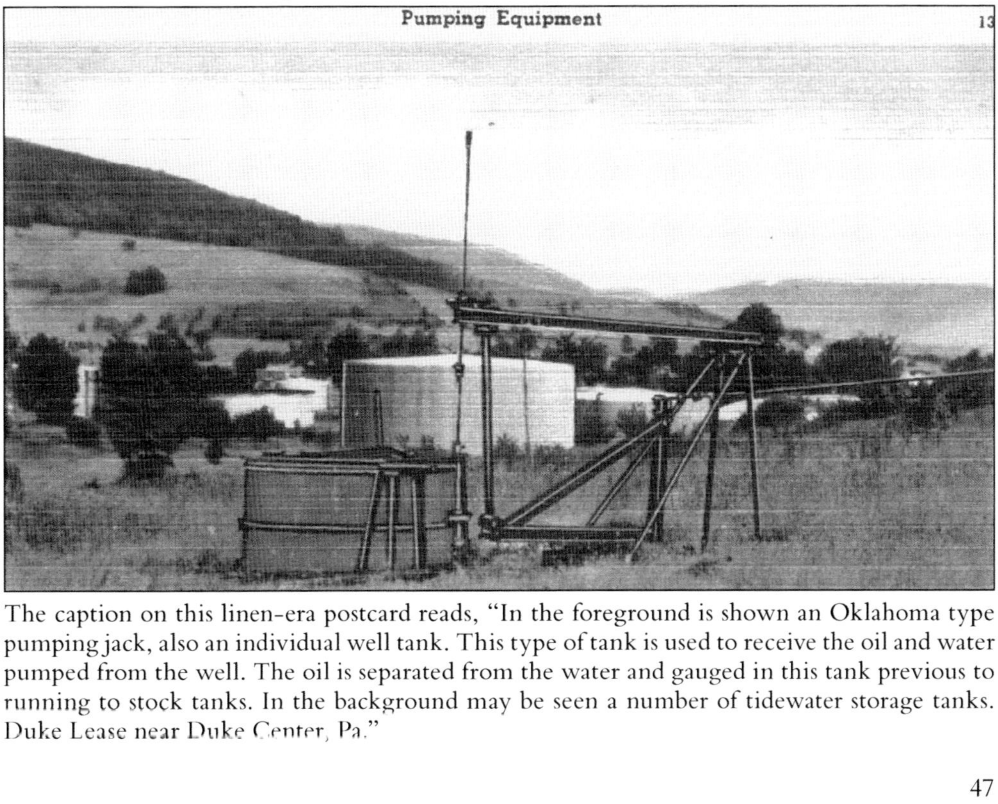

The caption on this linen-era postcard reads, "In the foreground is shown an Oklahoma type pumping jack, also an individual well tank. This type of tank is used to receive the oil and water pumped from the well. The oil is separated from the water and gauged in this tank previous to running to stock tanks. In the background may be seen a number of tidewater storage tanks. Duke Lease near Duke Center, Pa."

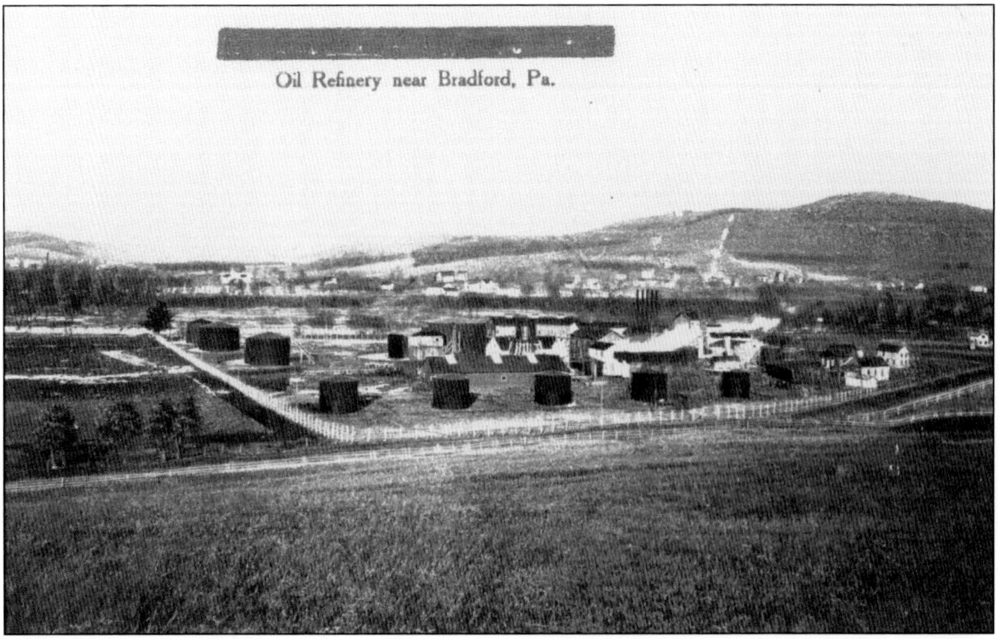

Bradford was a key oil production area but also was an important locale for refining petroleum. This early-1900s view of an "Oil Refinery Near Bradford" may be the view of the early Emery refinery.

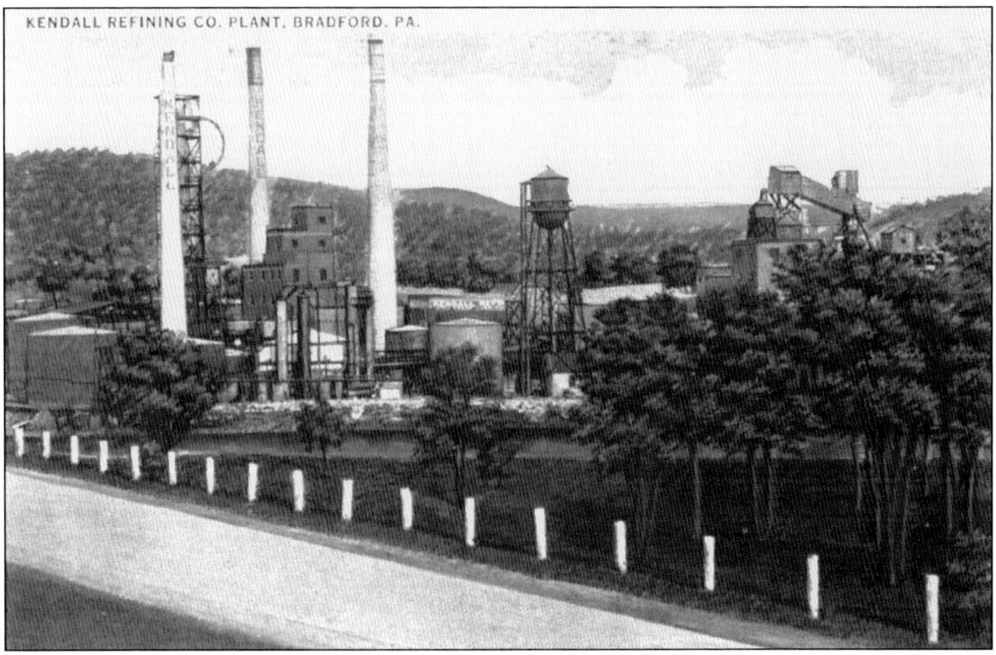

The Kendall or Bradford Oil Refinery had its origin in an earlier refinery built in Bradford by Lewis Emery Jr., an oilman who moved to Bradford from Oil City in the 1870s. The Bradford Oil Refinery was started along Kendall Creek in 1881 by independent oilmen Robert Childs, Eli Loomis, and William Willis. The Bradford Refinery is now the oldest continuously operating refinery in the United States, processing over three million barrels of crude annually.

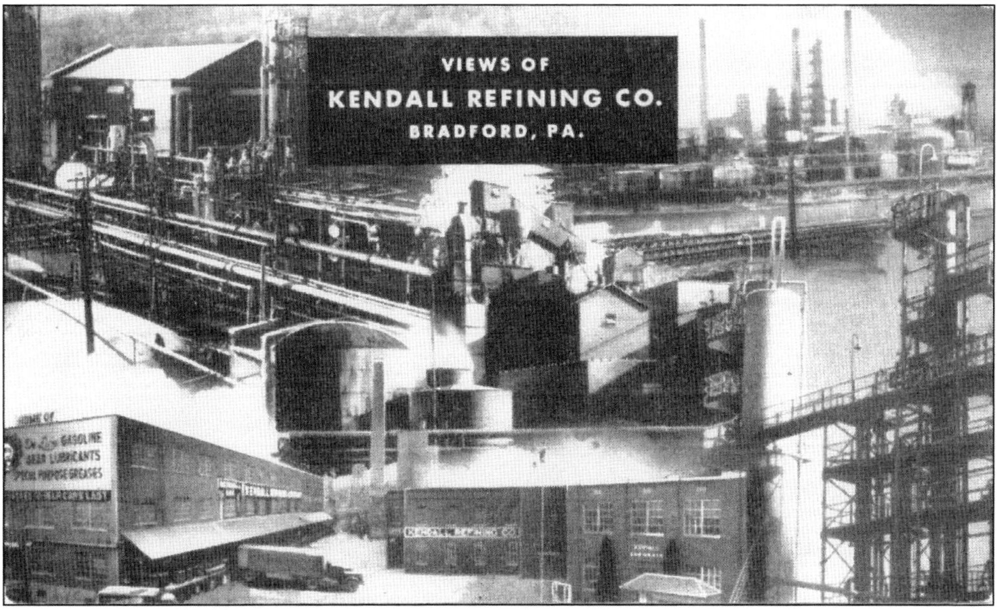

Here are further views of the Kendall Refining Company. The postcard above shows a variety of images of the refinery and its infrastructure, from production to distribution. The postcard below shows a linen-era view of the facility with the hills of the Tunungwant Valley in the background.

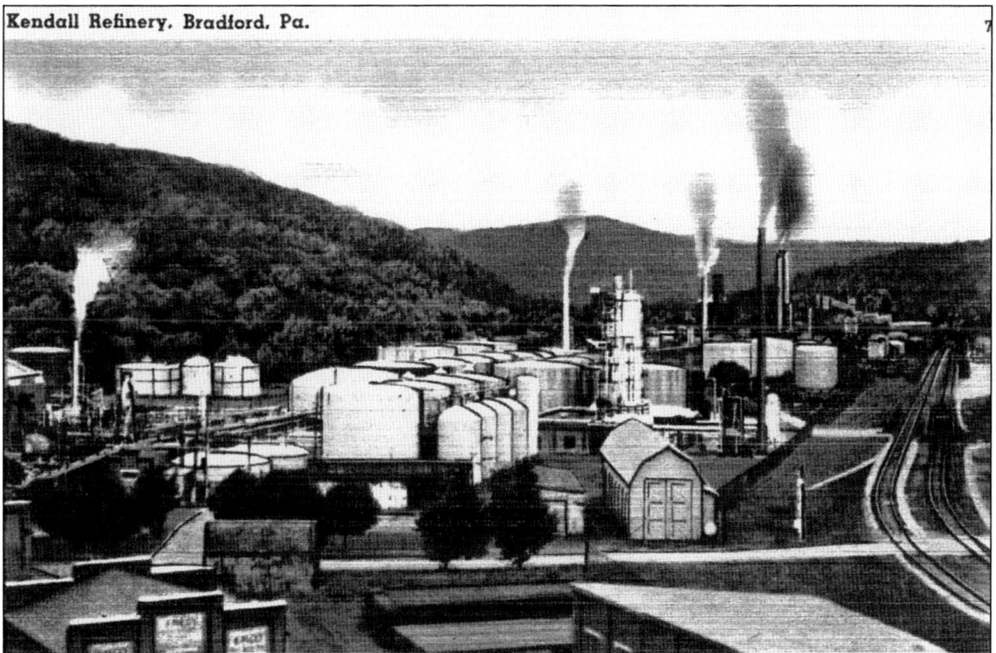

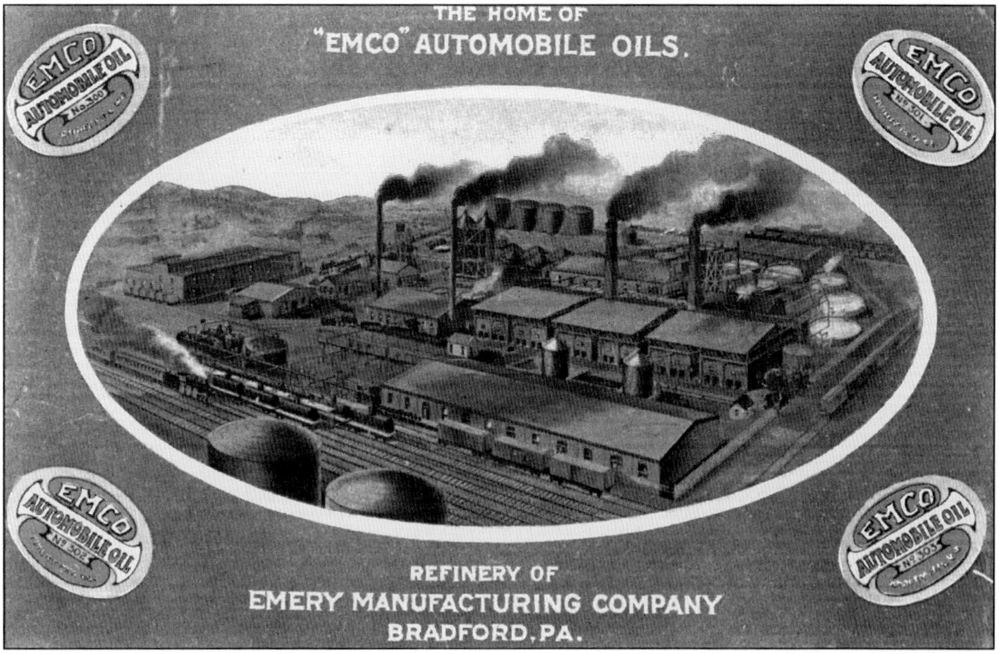

EMCO Automotive Oils were produced by the Emery Manufacturing Company of Bradford. The message on the back of this advertising postcard, postmarked 1911, reads, "We guarantee every gallon of oil sold by us to give entire satisfaction or it may be returned at our expense. We give you the privilege of using ten gallons from each barrel, for which no charge will be made, provided the oil will not meet your requirements."

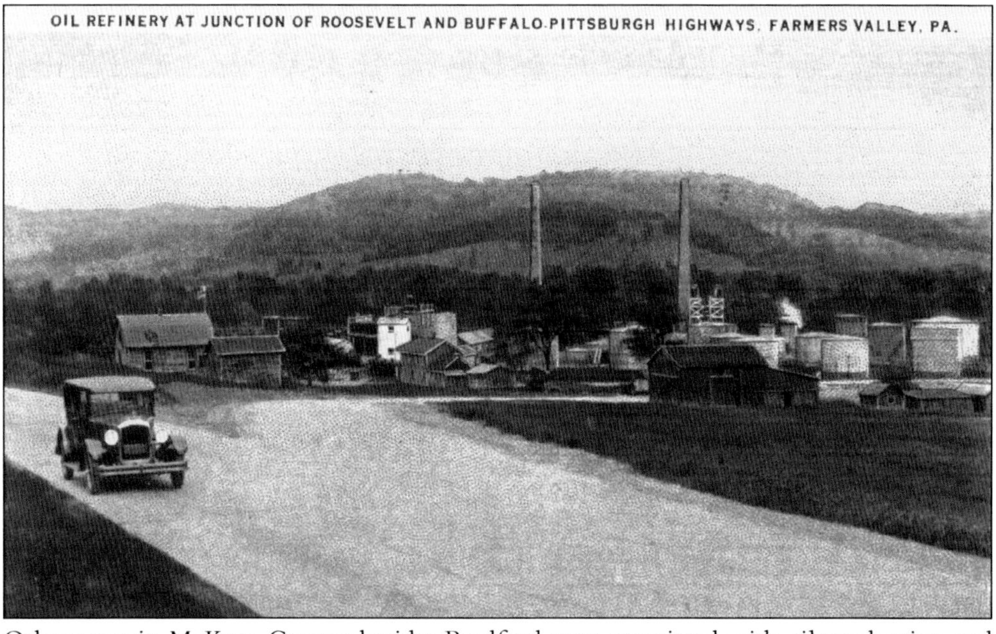

Other areas in McKean County besides Bradford were associated with oil production and refining. Farmers Valley was the site of the McKean Oil Refinery, which had 270 workers in the 1940s. A travelers guide at the time noted that Farmers Valley had a "thick pall of yellow smoke [hanging] over the plant . . . even on windy days."

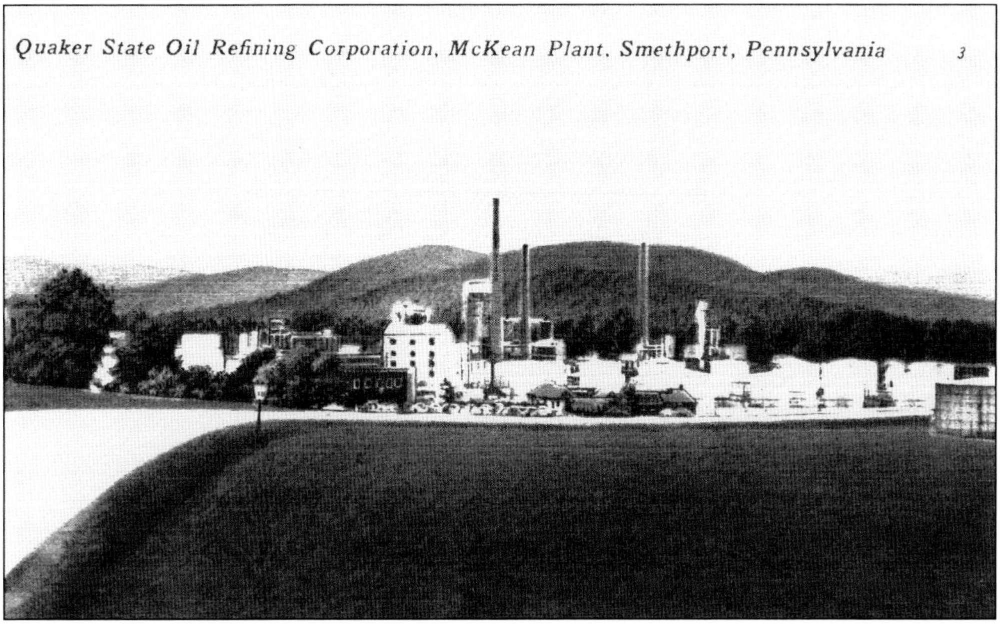

Smethport, McKean County, was the site of some interesting oil speculation in the 1860s; mine shafts were dug to 1,000 feet in an effort to reach oil. J. P. Morgan was rumored to have bankrolled this failed endeavor. This postcard shows the Quaker State refinery at Smethport.

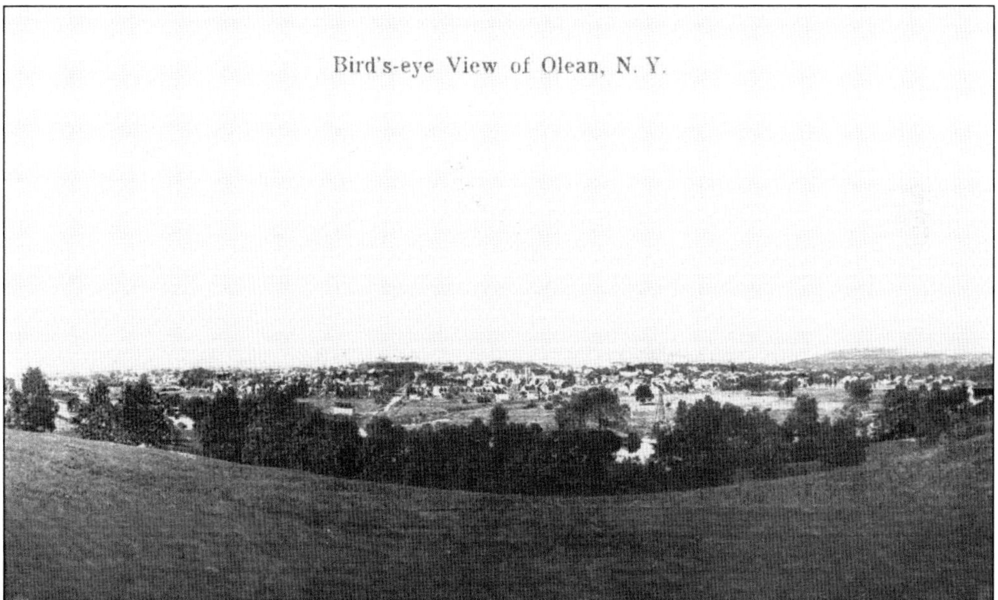

Olean, New York, was situated on the outskirts of the productive Bradford Oil Field. It was connected to the rest of the region by pipelines and railroads and became an important site for oil storage.

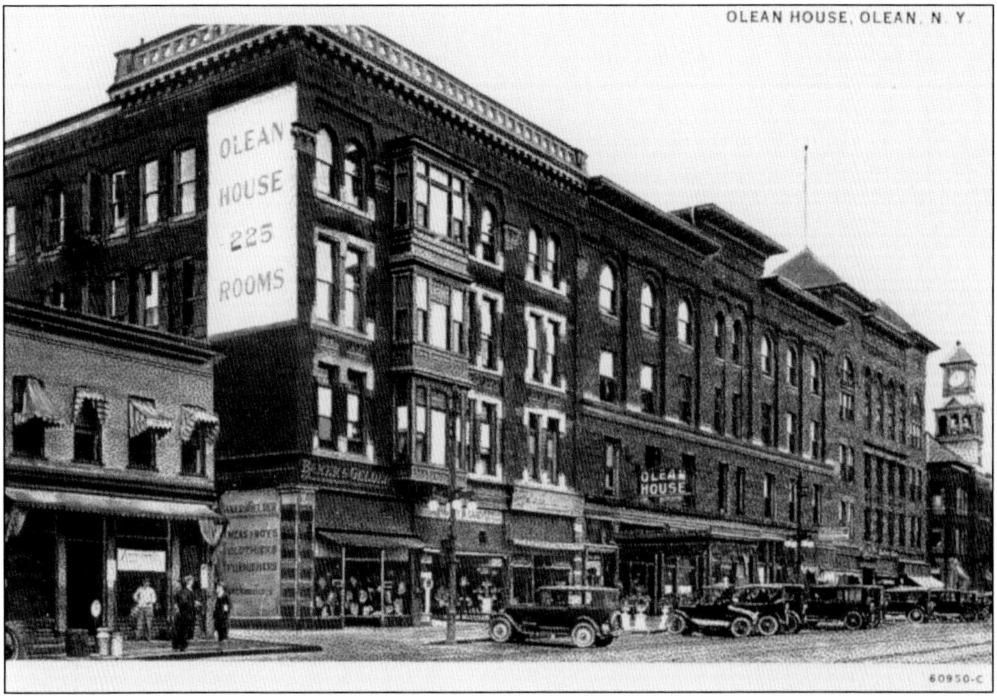

In the early 1900s, Olean had a population of nearly 17,000. Most of the local industry was associated with oil production and storage such as the Vacuum Oil Works and the Acme Oil Company's refinery that was relocated to Olean after the fire destroyed the refinery at Titusville. The postcard above from the 1920s shows the Olean House Hotel with 225 rooms. The postcard below, postmarked 1910, shows the extensive trolley lines that connected Olean with the Bradford Oil Field.

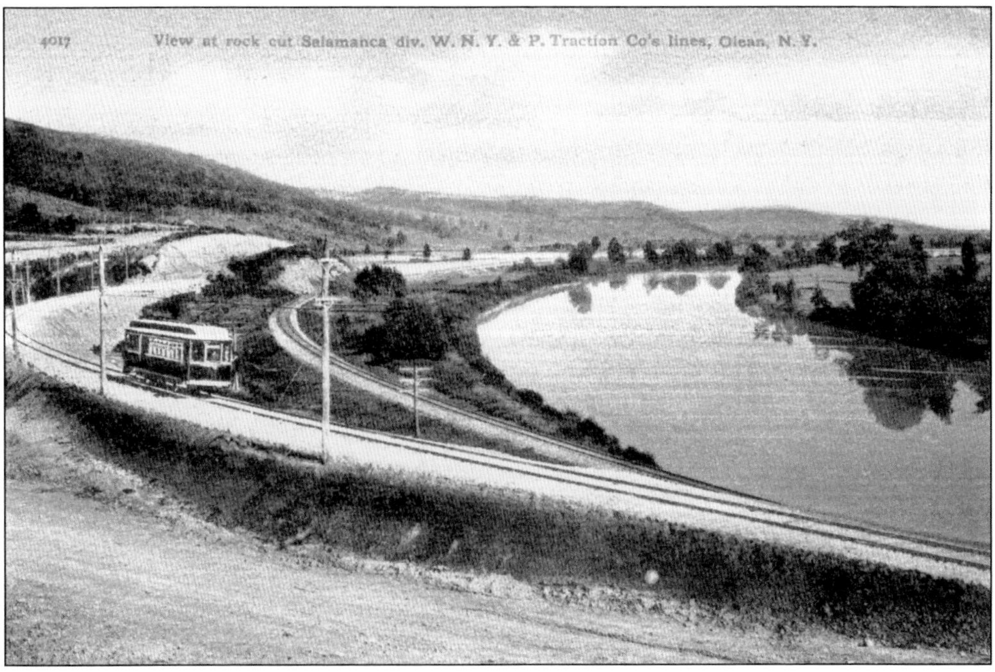

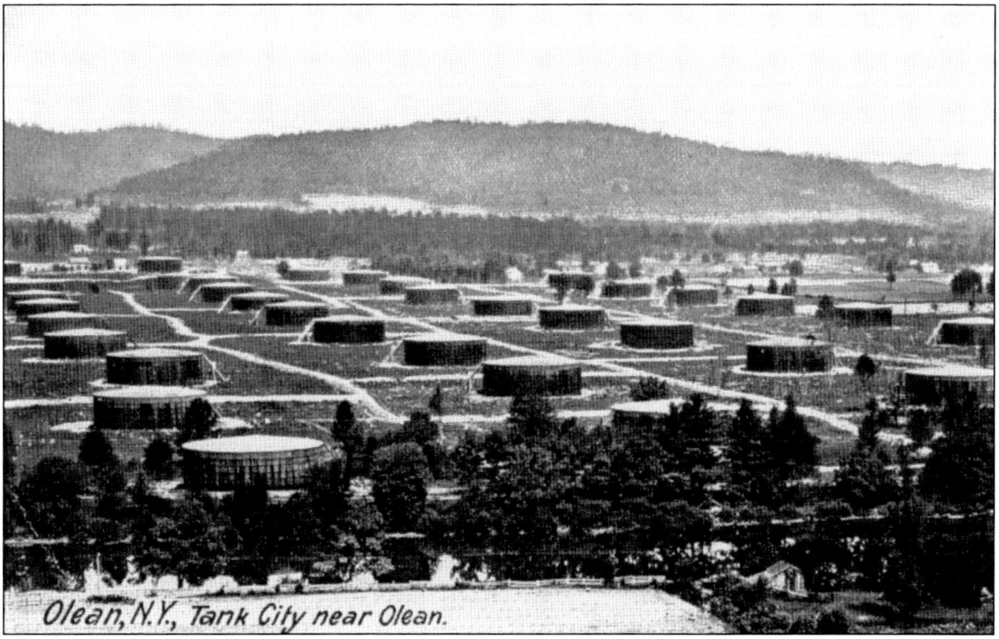

Olean was an important storage area for oil from the Bradford field. In the early 1900s, it had over 300 storage tanks clustered across the city with a capacity of nine million gallons. In 1881, John D. Rockefeller built a 315-mile pipeline from Olean to Bayonne, New Jersey, which operated until 1927.

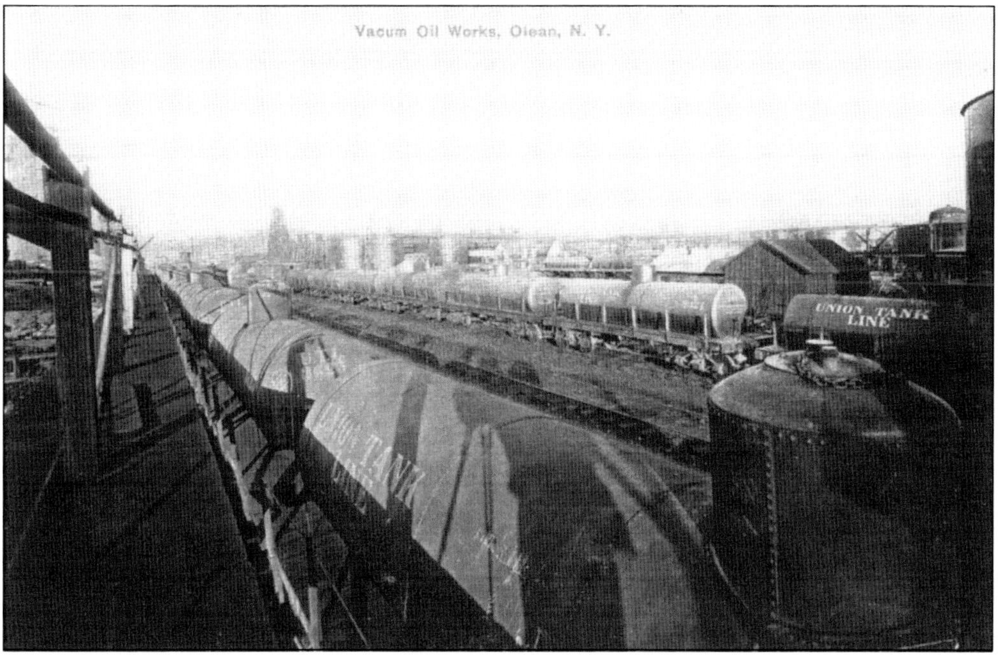

By the early 1900s, oil transport had developed beyond the days of teamsters hauling barrels and wooden tanking holding crude. The iron boiler railroad tank car was developed in 1869 and quickly became the industry standard. This postcard, postmarked in 1910, shows boiler-type railroad tank cars at the Vacuum Oil Works in Olean ready for transport.

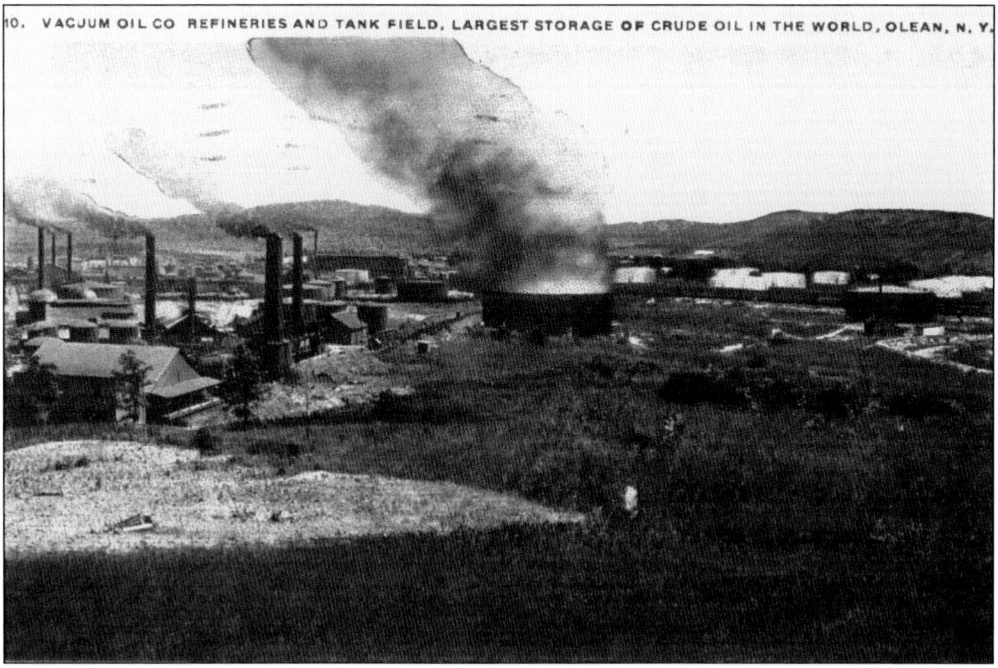

The Vacuum Oil Works had a commanding presence in Olean due to its shear size and capacity. The postcard above, postmarked in 1923, shows the "Vacuum Oil Co. Refineries and Tank Field, Largest Storage of Crude Oil in the World." On the postcard below, postmarked 1913, is a lament by the sender who appears to be bored with Olean: "All you can do in Olean, N.Y. is by [*sic*] a bag of peanuts and read the Evening herald."

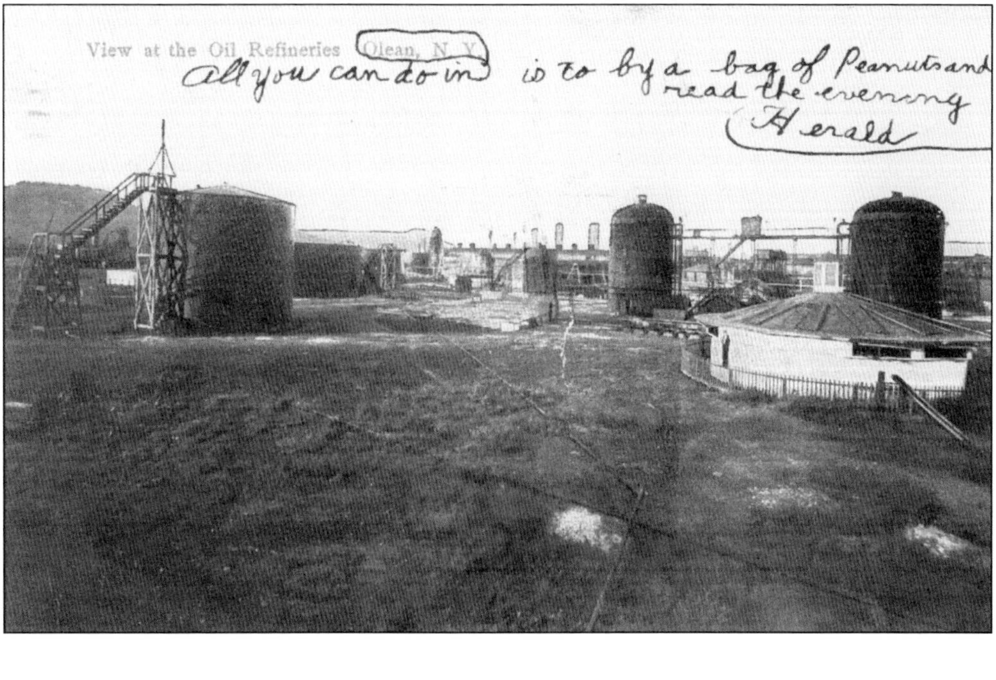

Four

Oil Speculation Moves South and East

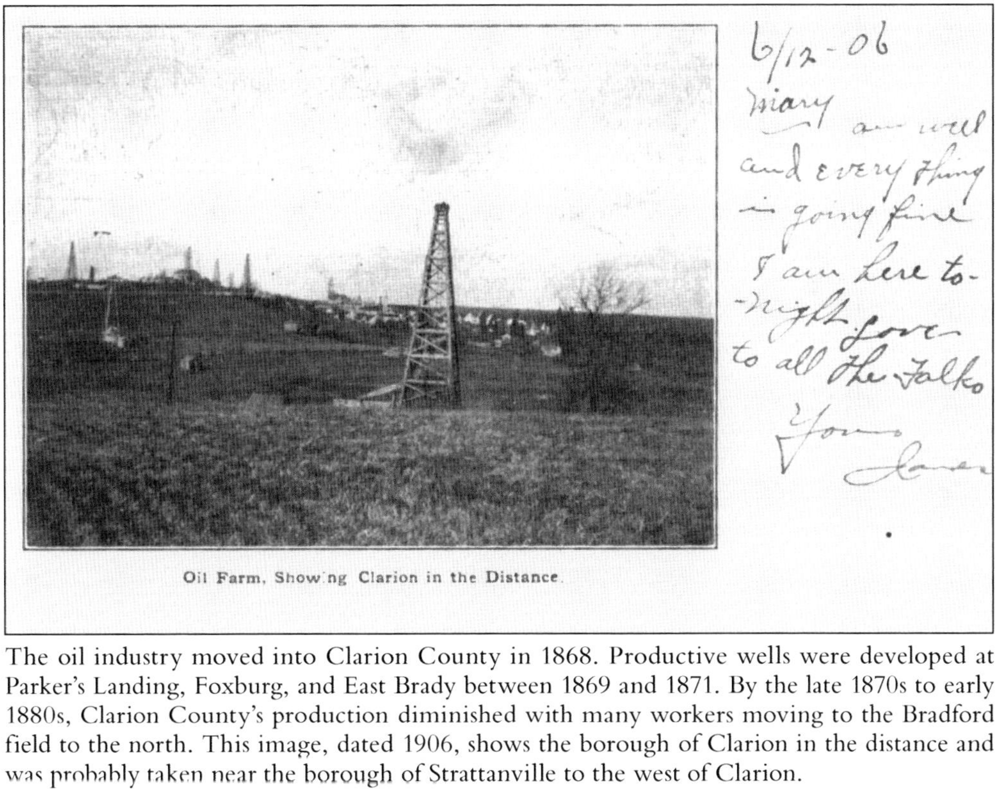

The oil industry moved into Clarion County in 1868. Productive wells were developed at Parker's Landing, Foxburg, and East Brady between 1869 and 1871. By the late 1870s to early 1880s, Clarion County's production diminished with many workers moving to the Bradford field to the north. This image, dated 1906, shows the borough of Clarion in the distance and was probably taken near the borough of Strattanville to the west of Clarion.

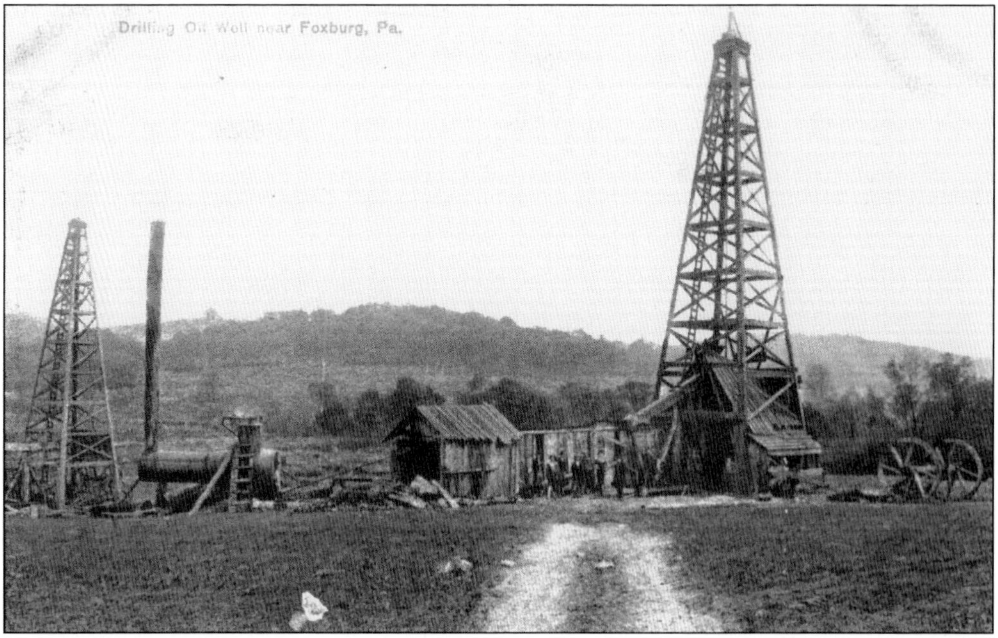

Foxburg, along the Allegheny River in Clarion County, was an important staging area for drillers and others who followed oil developments south in the late 1860s and early 1870s. The Mutual Pipeline Company, which supplied pipe to oil developments, had an office in Foxburg, and the town attracted oil speculators, brokers, and others with an interest in the industry. Foxburg became a destination, visited by Pres. Rutherford B. Hayes and Dom Pedro, emperor of Brazil, among other dignitaries.

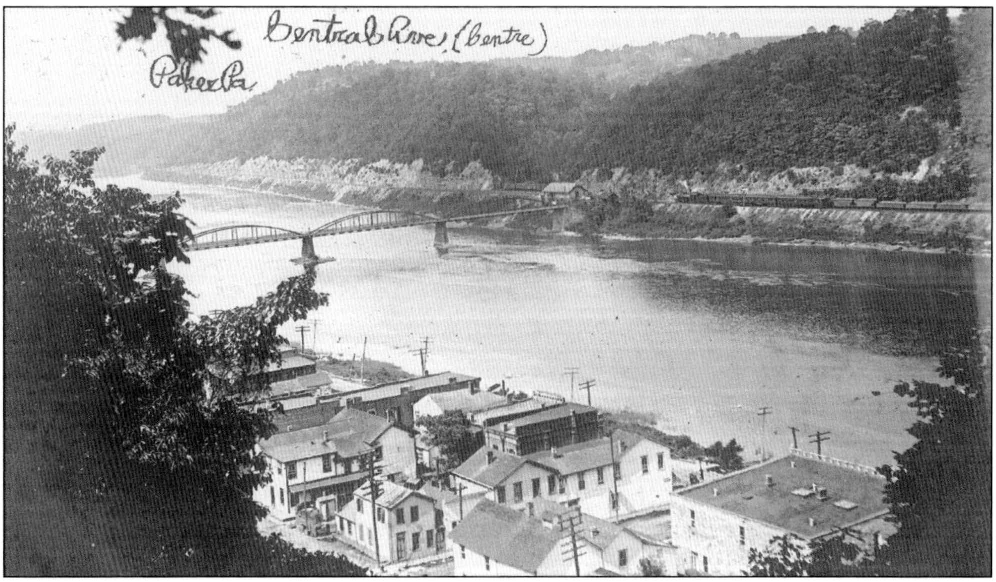

Parker, downstream of Foxburg in Armstrong County, was a quiet river town until the oil boom hit. The first oil well at Parker, Clarion No. 1, was drilled in 1865, yielding 18 barrels daily. Growth of the industry was slow, such that by July 1869 only 25 wells had been drilled, producing 310 barrels daily. But by November 1869, oil excitement struck and 1,058 wells had been drilled in the vicinity of Parker.

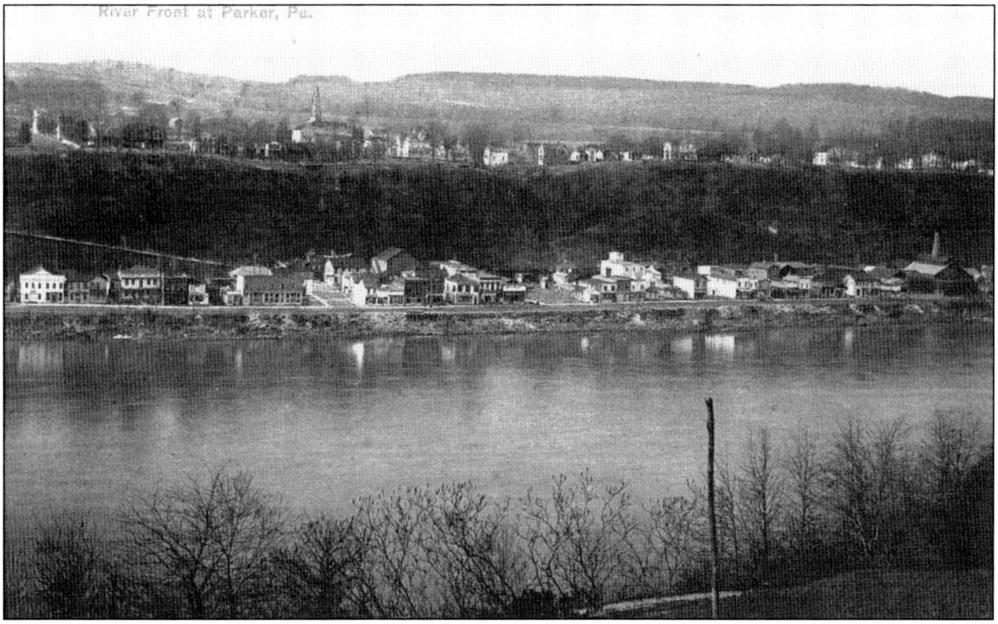

Parker became the hub of the lower oil fields with saloons, stores, hotels, and machine shops crowding both the bluff and riverfront portions of the town. An elevator provided transportation between the two levels of the town. Parker had an oil exchange with a $5,000 library and lavish club rooms. In 1873, with a population of 4,000, Parker was incorporated as a city. Parker's decline occurred quickly after the Bradford field opened. By 1880, only 1,000 inhabitants remained.

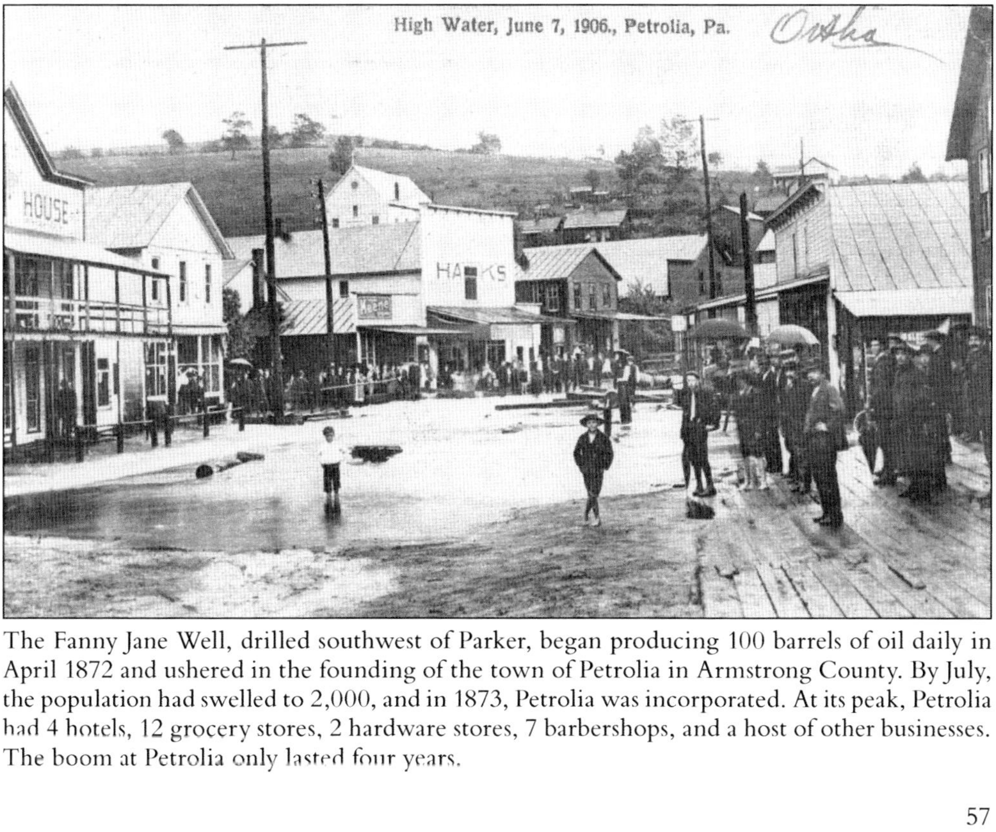

The Fanny Jane Well, drilled southwest of Parker, began producing 100 barrels of oil daily in April 1872 and ushered in the founding of the town of Petrolia in Armstrong County. By July, the population had swelled to 2,000, and in 1873, Petrolia was incorporated. At its peak, Petrolia had 4 hotels, 12 grocery stores, 2 hardware stores, 7 barbershops, and a host of other businesses. The boom at Petrolia only lasted four years.

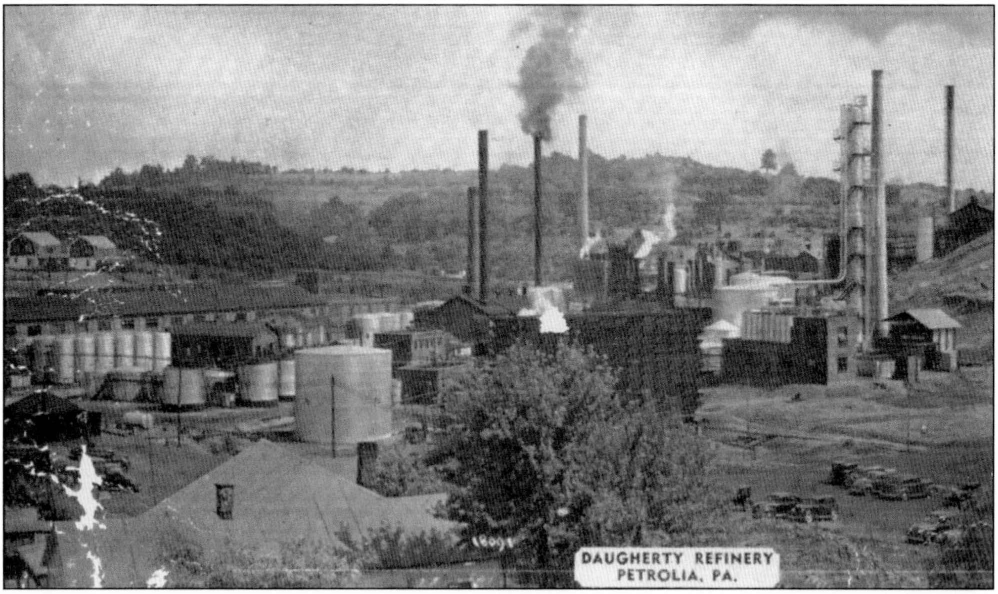

Its oil boom days over, Petrolia's population dropped to about 500 in the early 1900s. Oil refining became an important industry and remains so today. Two refineries were located around Petrolia, the Petrolia Refining Company and the Daugherty Refinery, shown in this Silvercraft postcard from the 1940s. Products from these refineries included petrolatum and lubricating and illuminating oils.

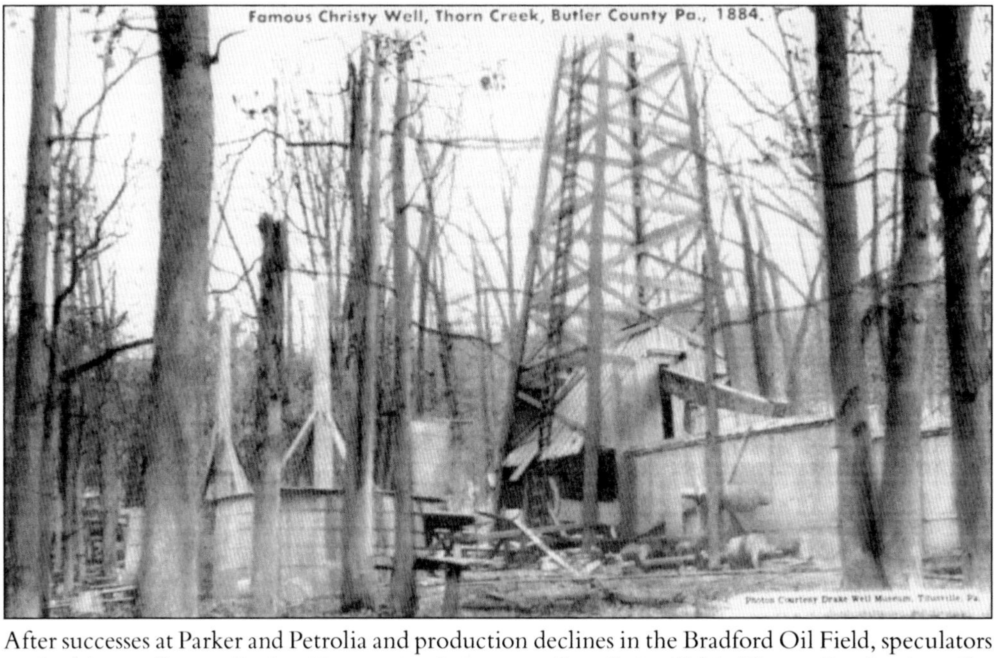

After successes at Parker and Petrolia and production declines in the Bradford Oil Field, speculators pushed into adjacent Butler County in their quest for oil. In 1884, Thorn Creek, six miles southwest of Butler, became the next boom field. The Phillips Well on Thorn Creek, drilled in August, produced 500 barrels daily. The operators drilled 10 feet deeper on September 15, and the well produced 4,200 barrels in one day. The nearby Christy Well produced 5,000 barrels on its first day of production.

Not all Butler County wells were highly productive. The Spotty McBride Well, seen in this image, was drilled near Butler in the early 1900s. It was a gusher at first, signaling good prospects for the oil pool in the vicinity. Thirty wells were sunk near the Spotty McBride Well, with less than a third of the wells producing oil at a total of 50 barrels daily, 15 of those contributed by the Spotty McBride Well.

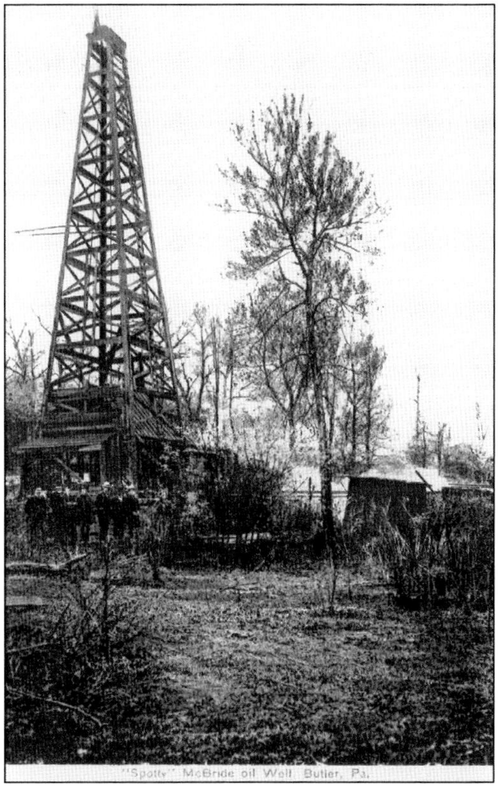

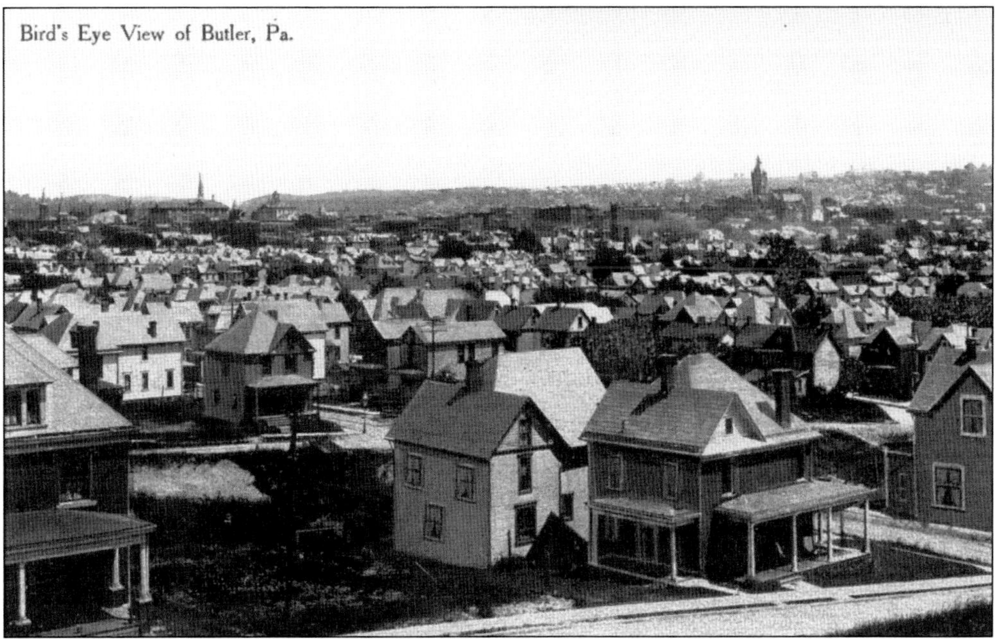

Butler was established in 1803. In the 1940s, Butler was an important industrial town with over 7,000 workers engaged in the manufacture of plate glass, refrigerators, railroad cars, metal products, and of course, machinery and supplies for the oil industry. This bird's-eye view of Butler is dated 1910.

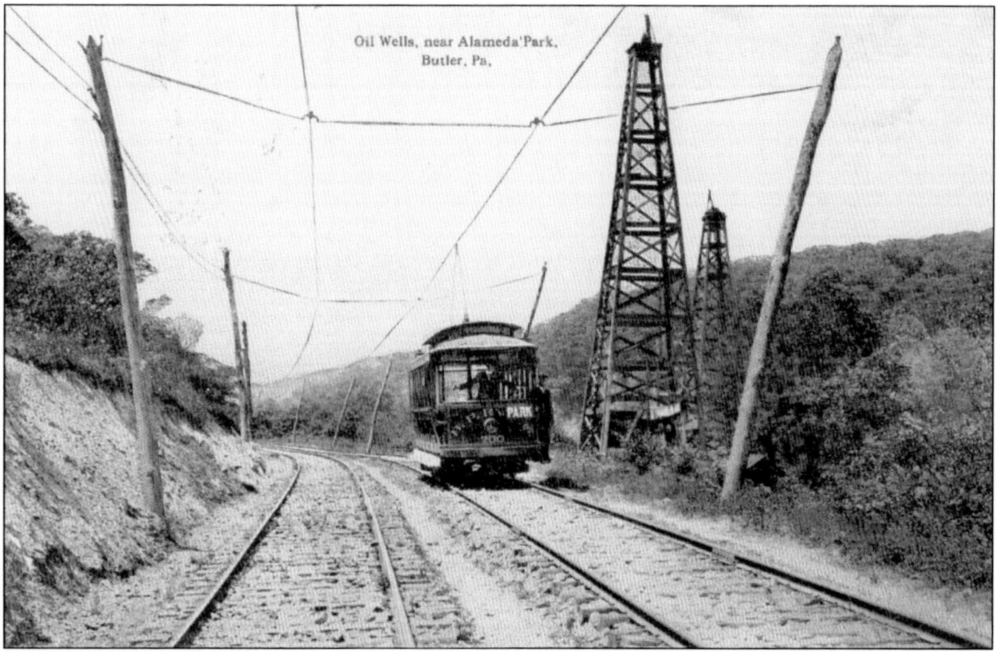

This postcard image from the early 1900s shows two of the important factors in the growth of Butler—oil and railroads. This postcard shows a view of "Oil wells near Alameda Park, Butler." Alameda Park was a popular destination for residents of the region and beyond, offering many popular amusements.

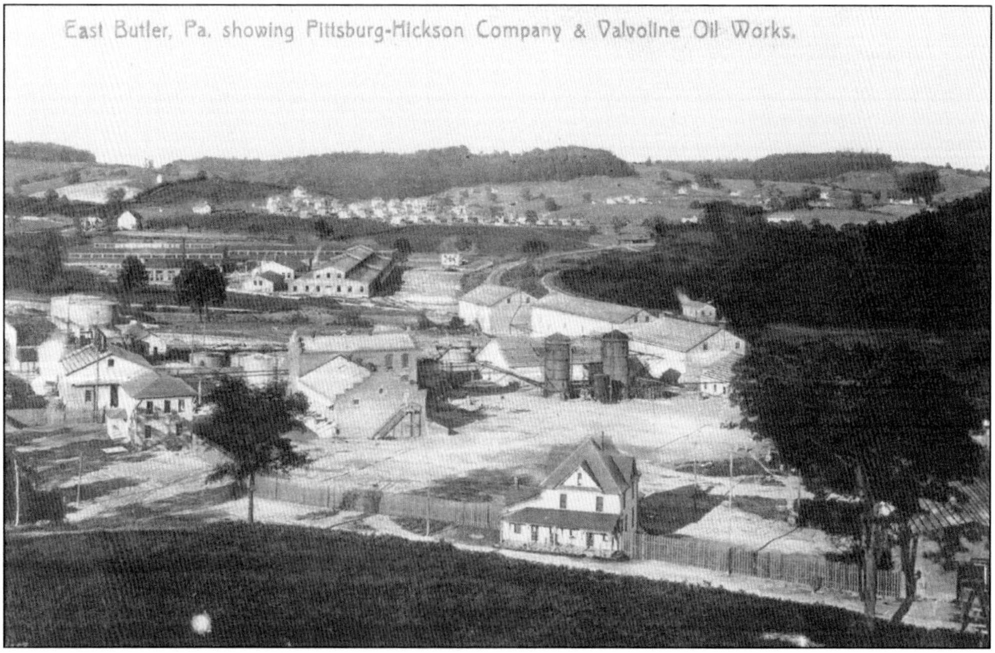

East Butler was an industrial hub in the greater Butler area. This early-1900s view of the town shows the Pittsburgh-Hickson Company and the Valvoline Oil Works.

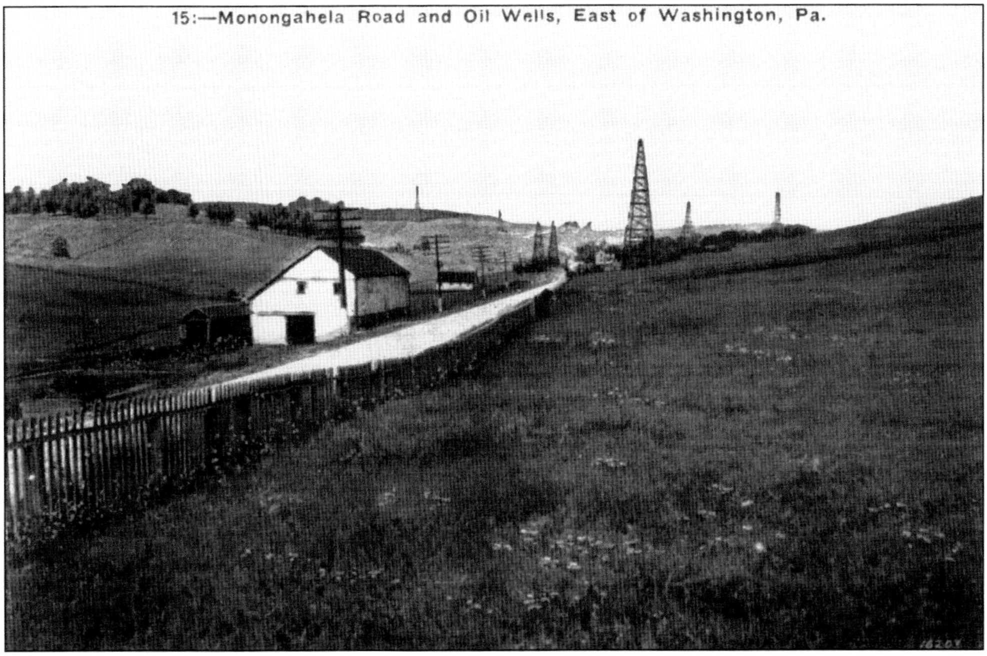

In the mid- to late 1880s, oil production in Pennsylvania began to shift to the southwest. Washington County, at the extreme southwestern portion of the state, had both shallow oil and gas fields. Washington County's oil fields came into production from 1885 to 1887. The postcard above shows a view of the "Monongahela Road and Oil Wells, East of Washington, Pa." The postcard at right, postmarked 1911, shows a "Flowing Oil Well, Washington, Pa."

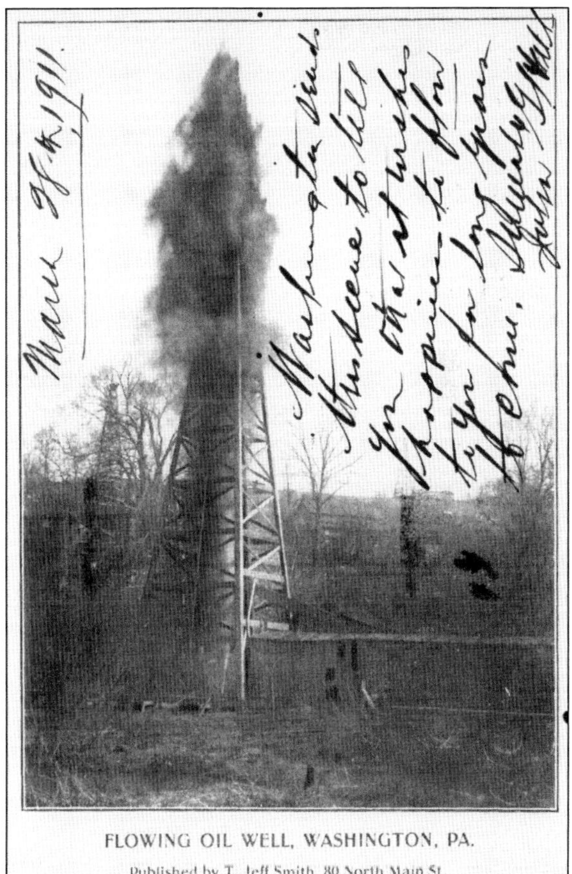

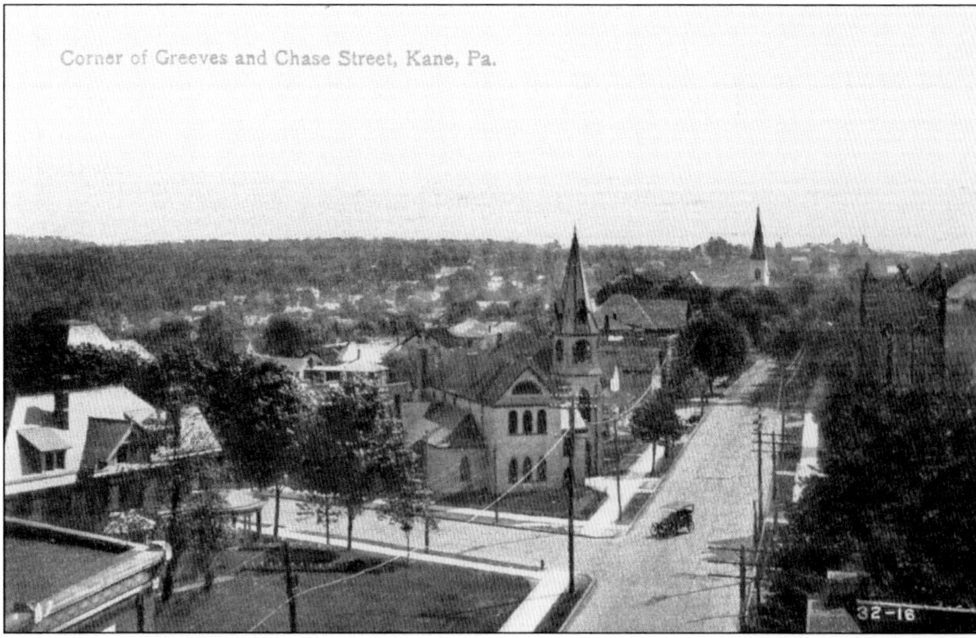

Kane had a population of about 8,000 in the early 1900s. Industries at the time included four window glass plants, two bottle glass plants, and brush handle, cutlery, and blind factories. Originally a lumber town, oil and gas production became a mainstay in the region in 1885 when the oil fields around Kane were developed. This postcard shows a bird's-eye view of the corner of Greeves and Chase Streets in Kane.

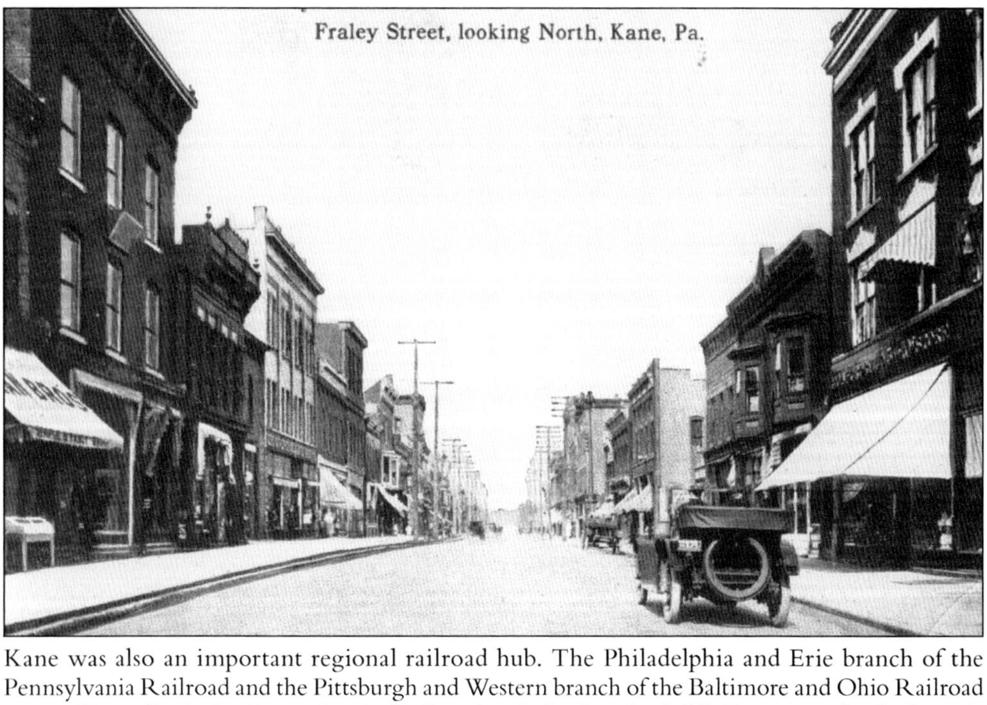

Kane was also an important regional railroad hub. The Philadelphia and Erie branch of the Pennsylvania Railroad and the Pittsburgh and Western branch of the Baltimore and Ohio Railroad met at Kane. Both the Pennsylvania Railroad and the Standard Oil Company also had repair shops at Kane. This view of Fraley Street looking north in Kane was postmarked in 1923.

This split-image postcard, postmarked 1915, shows the depots of the two main rail lines that met at Kane. The top image shows the Pennsylvania Railroad depot. The bottom image shows the Baltimore and Ohio Railroad depot. Kane was also the eastern terminus of the Kane and Elk Railroad in the early 1900s. The Kane and Elk Railroad was an important regional lumber rail line.

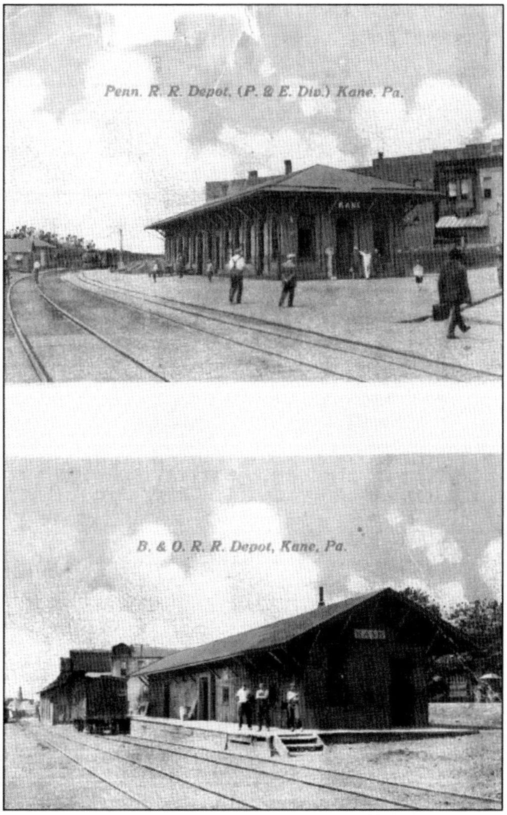

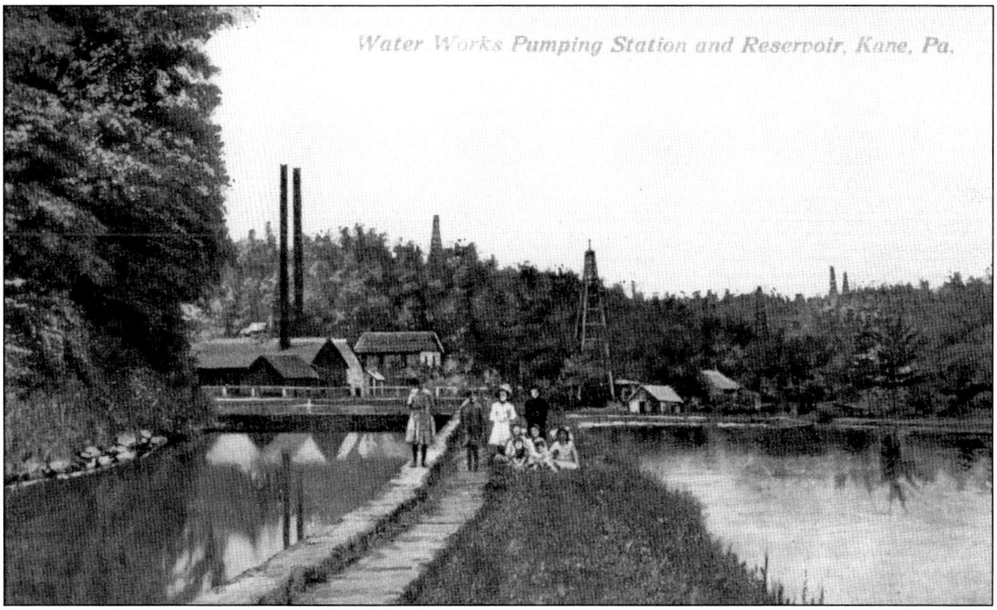

This early-1900s view of the waterworks pumping station and reservoir in Kane shows derricks standing on the shore of the reservoir. At the time, Kane was supplied with water by the Spring Water Company, whose supply included well, spring, and surface waters. The water was considered of good quality except during dry seasons.

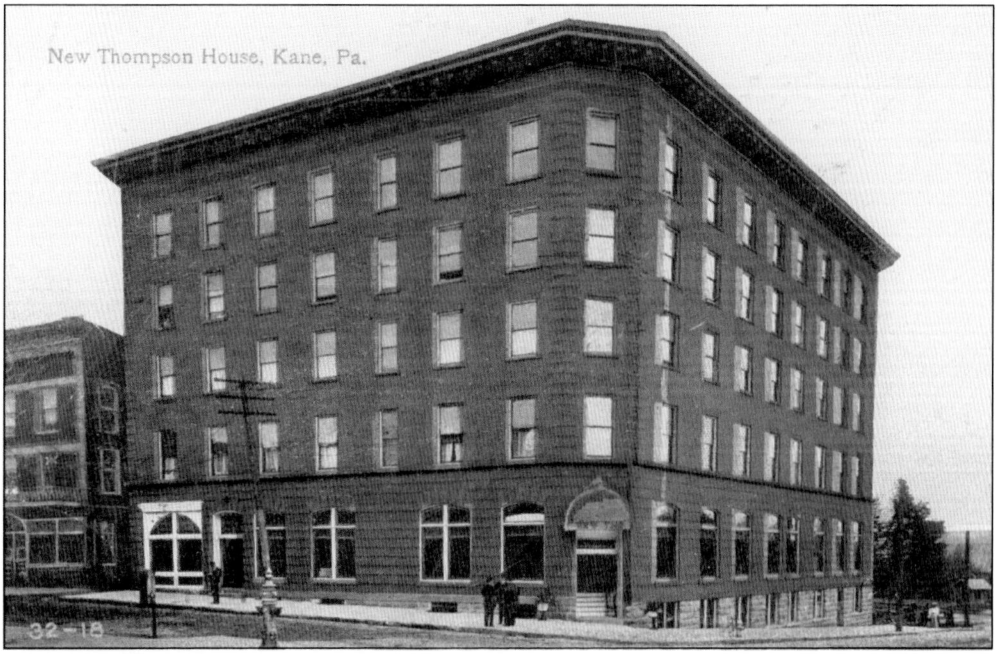

New Thompson House, located on Greeves Street in Kane, was, and still is, one of the most prominent buildings in the town. It served duty as a hospital in the early 1900s when a laundry was added to the bottom floor. This view of the New Thompson House was postmarked in 1915.

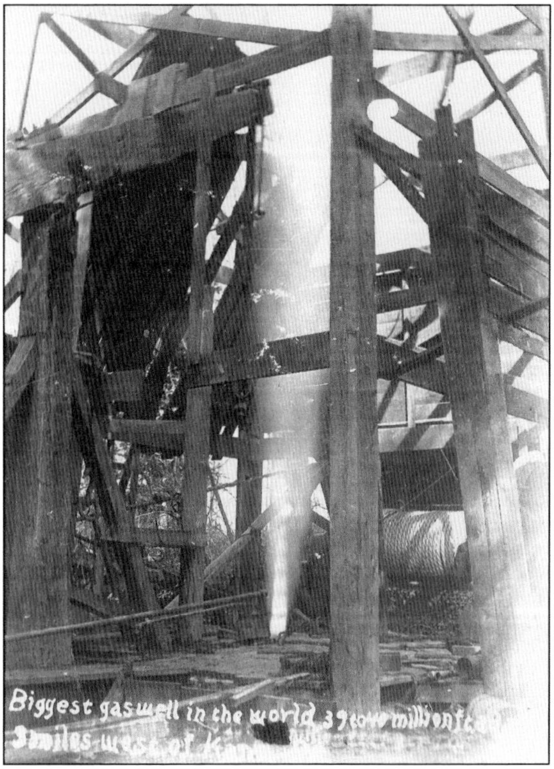

Kane lies atop both shallow gas and oil fields, and early oil speculators often had to contend with waste natural gas until a collecting and marketing infrastructure developed for the product. This postcard from the early 1900s shows a blowing gas well near Kane. The caption reads, "Biggest gas well in the world 39-30 million ft. a day 3 miles west of Kane."

Five

Titusville, Oildom's Queen City

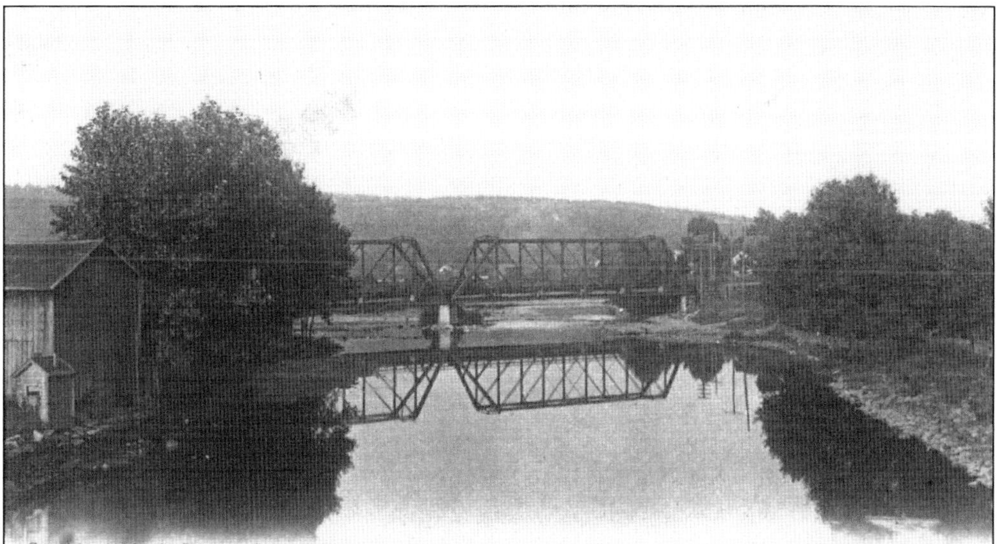

Titusville owes its beginning to Jonathan Titus, who settled in the area in 1796. Originally named Edinburg, Titusville was incorporated as a borough in 1847. Lumbering was the primary business. When Edwin Drake first visited Titusville in 1857, the town had about 125 residents, two hotels, and no churches. By 1870, after oil was struck nearby, the population had swelled to nearly 12,000, and Titusville became the "Queen City" of the oil region.

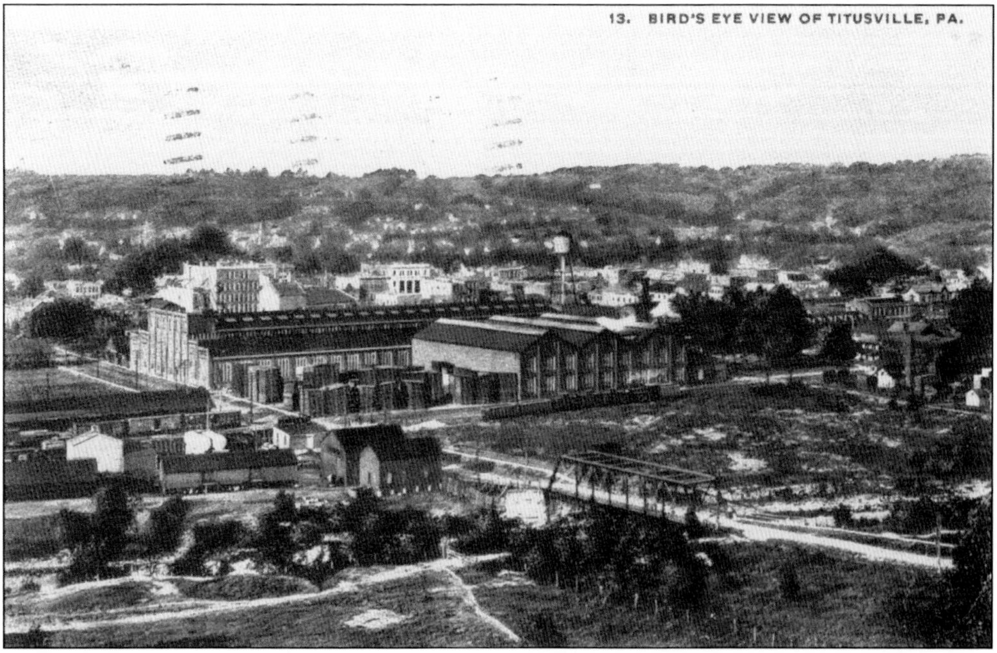

Although built and sustained by oil, Titusville's industries were diverse and extensive by the early 1900s. At the time, three oil refineries, several steelworks and iron foundries, a paraffin works, a soap works, the Queen City Tannery, and the Grasselli Chemical Company were important industries. This bird's-eye view of Titusville was postmarked in 1924.

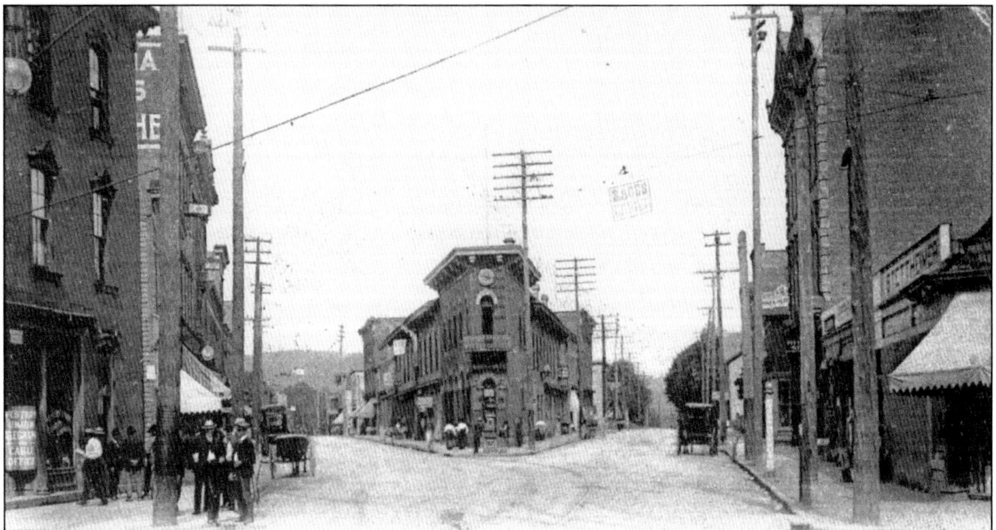

The business district of Titusville, with its many stores and banks, was largely located along Central Avenue, Spring Street, and Diamond Street. This postcard of Spring and Diamond Streets was postmarked in 1905. At the right of the image is Stettheimer's Clothing Store, which was located on the Hershberg Block. Stettheimer's was a men's clothing store that sold ready-made clothing and other dry goods.

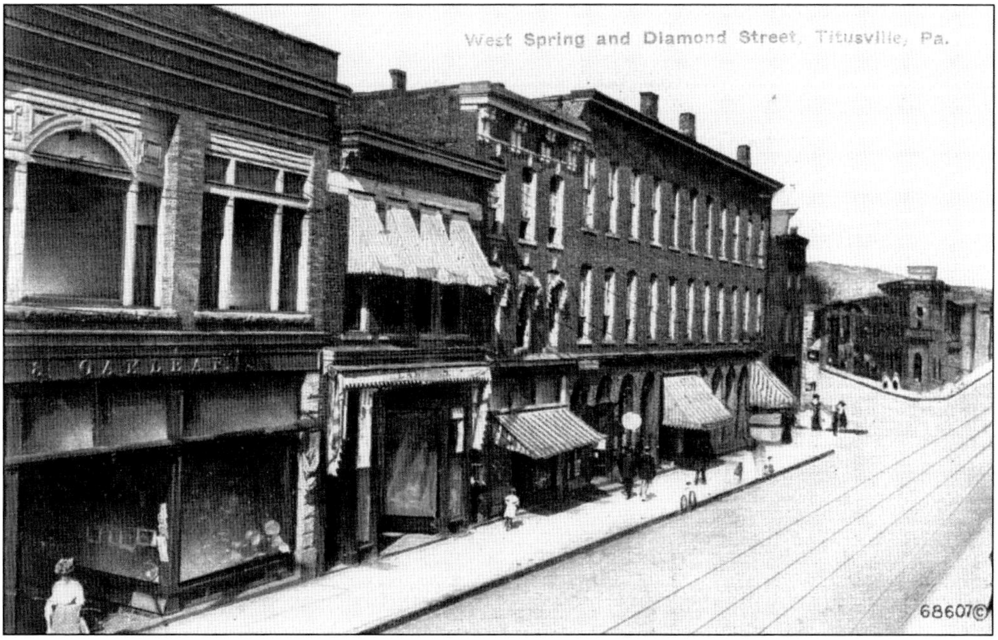

The view of "West Spring and Diamond Street" in Titusville was postmarked in 1911. At the left of this image is the Cohn and Oakleaf storefront. The Cohn and Oakleaf Company was a retail stationery seller and bookstore. It also published and sold postcard images of Titusville such as this one.

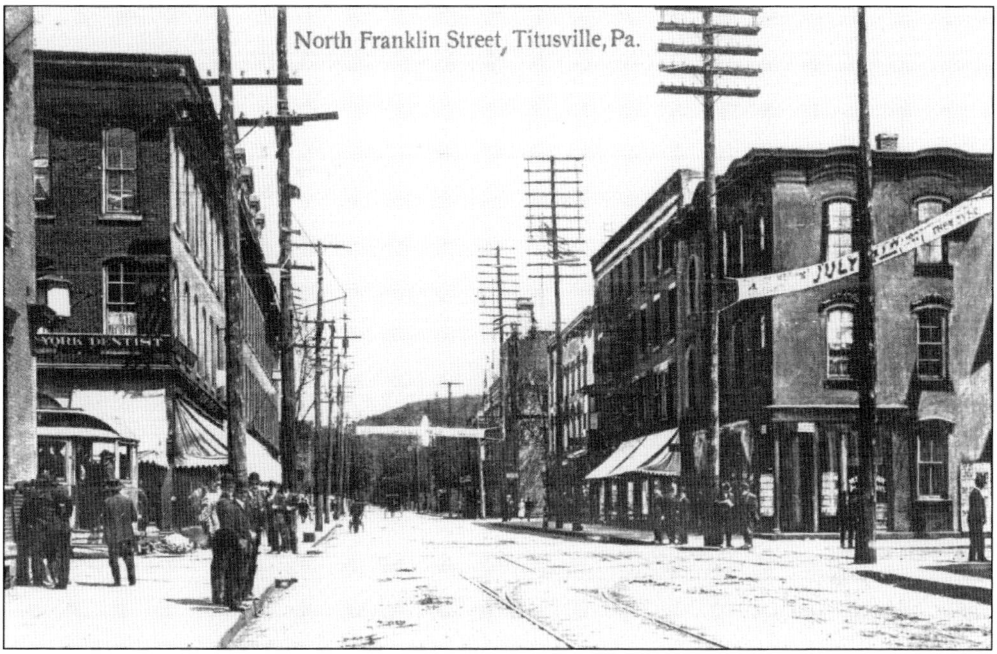

This postcard, postmarked 1908, shows a view of North Franklin Street in Titusville. Banners draped across the streets advertise a Fourth of July celebration. At the left of this image is the Chase-Stewart Block, started in 1870 and completed a few years later. At the corner on the right of the image is the First National Bank built in 1864.

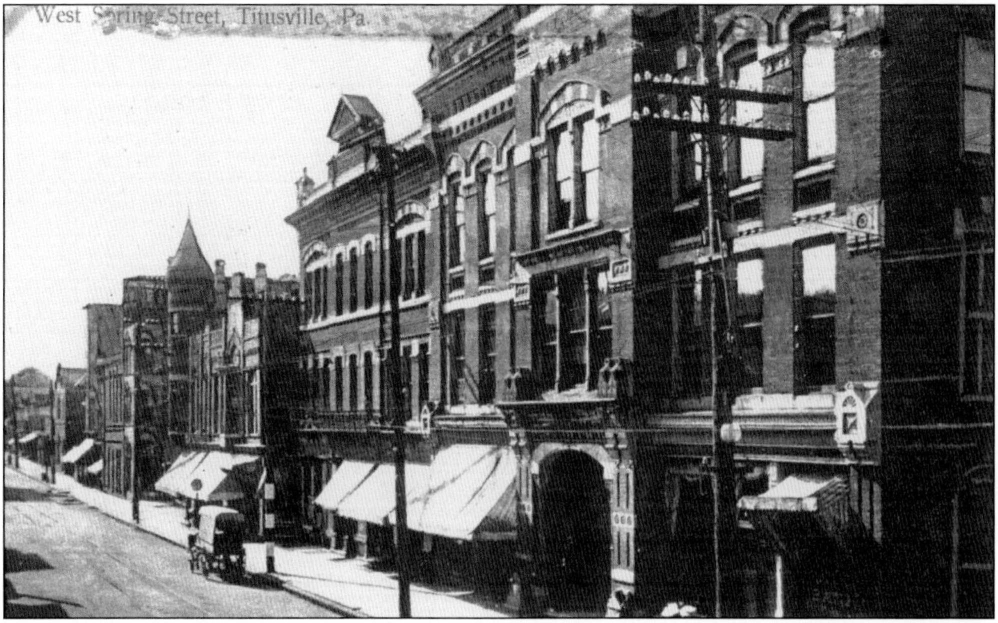

As the oil industry grew, oil exchanges, organizations in which buyers and sellers could meet and conduct business, were established in many of the oil region's hubs. In January 1871, over 100 businessmen organized the Titusville Oil Exchange. As the exchange grew, there was interest in erecting a building that would epitomize Titusville's prominence in the oil market. The new Titusville Oil Exchange building, seen in this postcard postmarked 1908, was dedicated on January 27, 1881.

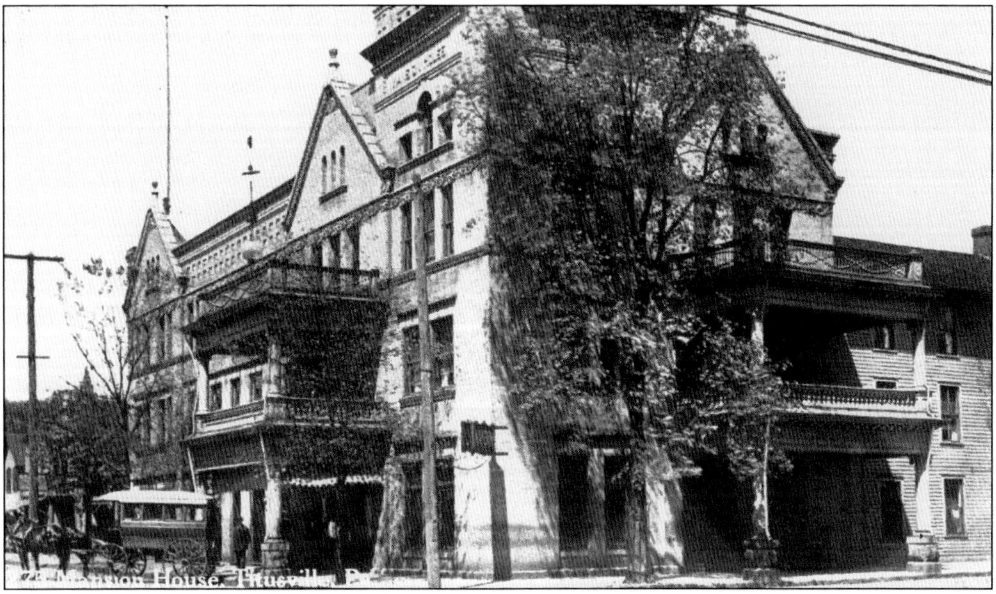

The Mansion House had its origins in 1864, when it was a three-story wood frame building. After numerous owners, O. E. Gleason and L. M. Lockwood rebuilt and expanded the hotel in 1897. The old wood frame building was brought down in stages and replaced by a yellow brick Victorian Romanesque hotel, considered "one of the finest and best appointed" in the oil region. In 1946, the Mansion House was remodeled and renamed the Colonel Drake Hotel.

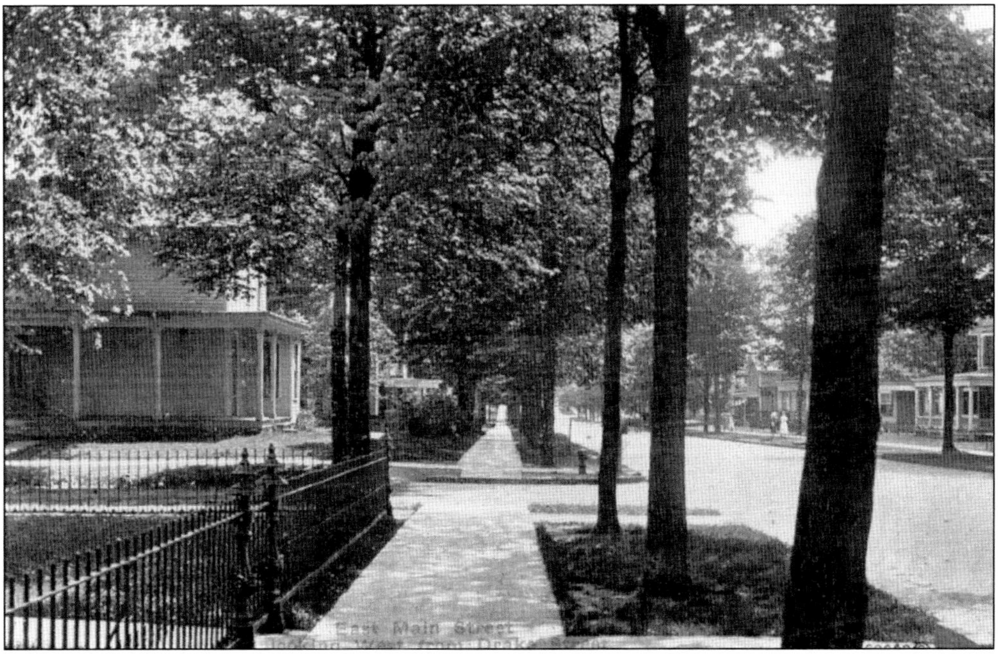

Titusville earned its nickname, the "Queen City," because of its fine mix of industry and culture, attracting many to reside in the town. Main Street, seen in this postcard postmarked 1912, was perhaps Titusville's most important and attractive residential area. Main Street included among its residents John A. Mather, oil field photographer; Franklin S. Tarbell, designer of the first wooden oil storage tank; and Calvin M. Payne, manager of the National Transit Company.

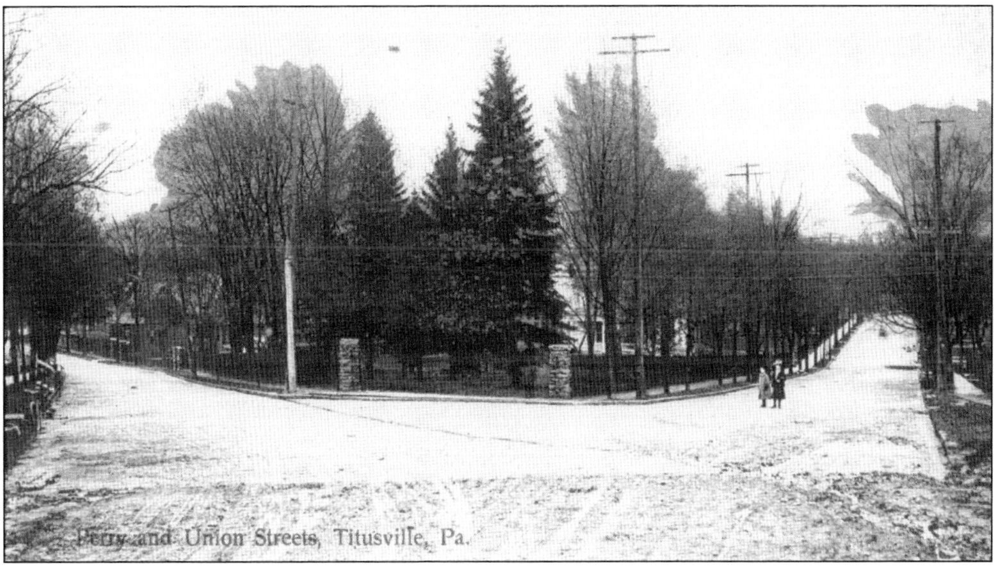

Oil wealth was readily apparent in Titusville in the form of ornate Victorian mansions and large residences located across the town. This postcard, postmarked 1908, shows the corner of Perry and Union Streets in Titusville, where many influential people in town resided. Oilmen such as Byron Benson, who organized an independent pipeline company that successfully challenged the Standard Oil Company's oil transport monopoly, and James C. McKinney, who once controlled over 100,000 acres of oil lands, had mansions nearby.

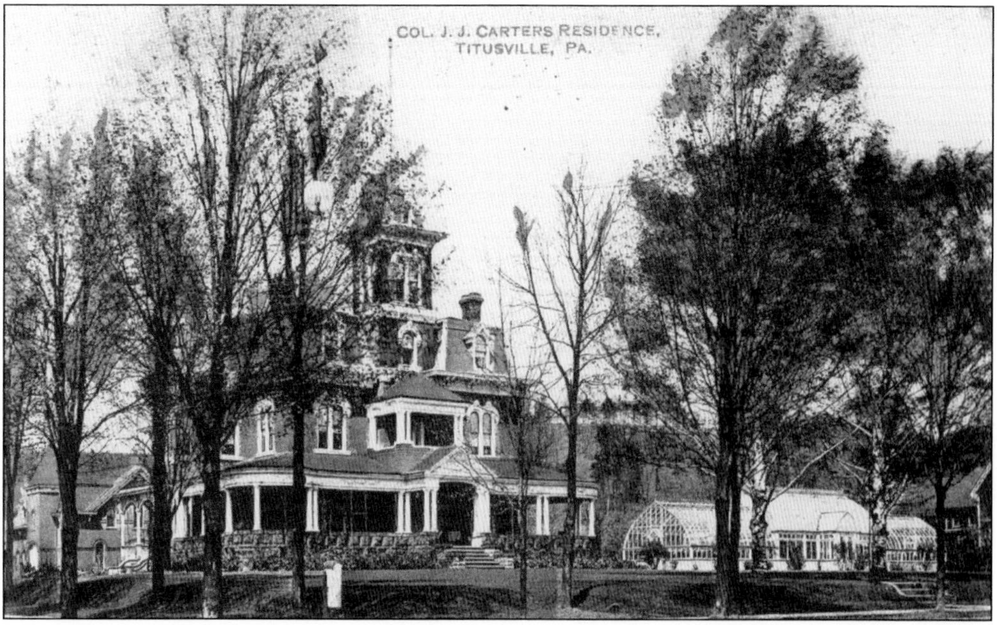

The Col. J. J. Carter residence was built by John C. Bryant, an Irish iron founder, in 1871. John J. Carter bought it in 1884. Carter got his start with a well in Pleasantville and then moved to the Bradford field, where his wells produced over 1.5 million barrels of oil in 20 years. Louis McKinney bought the house in 1926 and extensively remodeled it. Today it serves as the administration building for the University of Pittsburgh at Titusville.

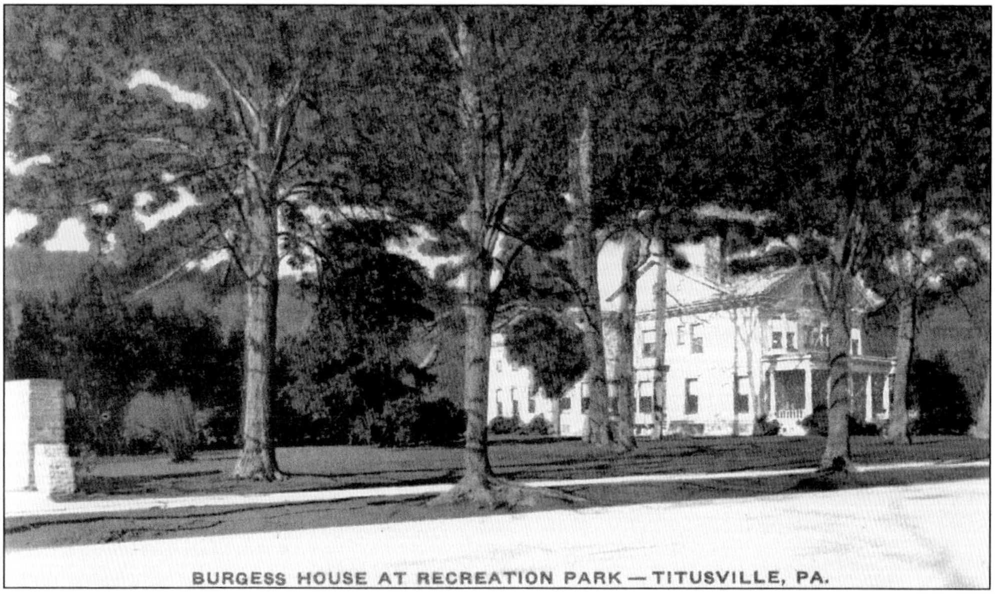

Jonathan Watson moved to Titusville in 1845 to pursue an interest in the lumber industry. After oil was struck, Watson began to lease oil lands and would become the oil region's first millionaire. In 1870, he built this estate for $100,000. By 1878, Watson had fallen on hard times and died penniless. After duty as a sanatorium, the estate was bought by Charles Burgess, owner of the Cyclops Steel Works, in 1896. The estate is now part of the Titusville Community Center at Burgess Park.

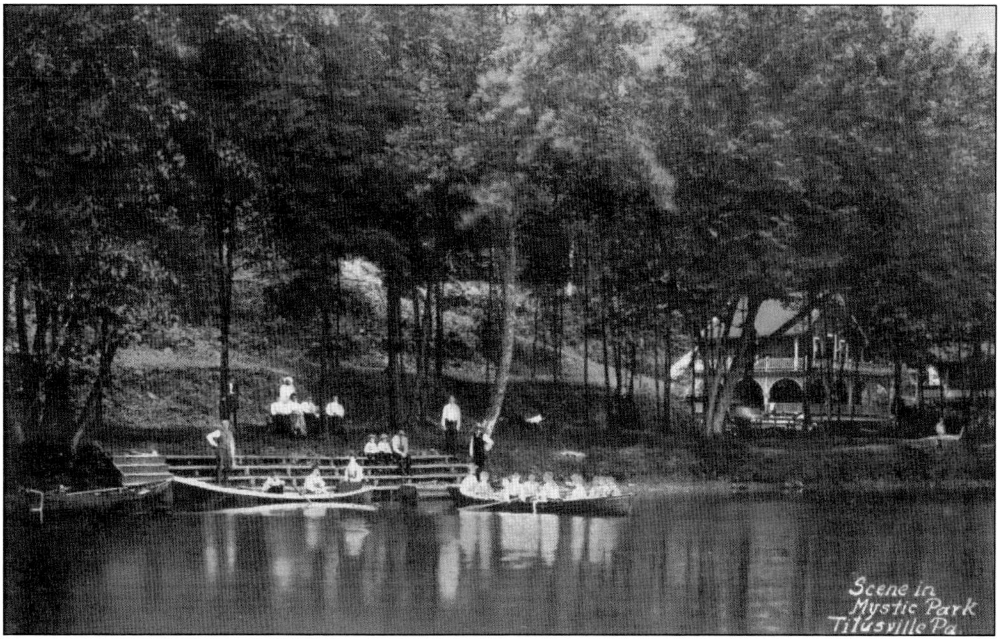

Titusville, like many larger communities of the time, had a trolley line that linked the town to others in the region. The Titusville Traction Company was organized in 1897 and by 1898 had built a trolley line from Titusville to Pleasantville. The line was extended to Hydetown, north of Titusville, in 1901. Mystic Park, seen in this postcard postmarked 1914, was built at Hydetown to attract trolley riders. The trolley line in Titusville closed in 1924.

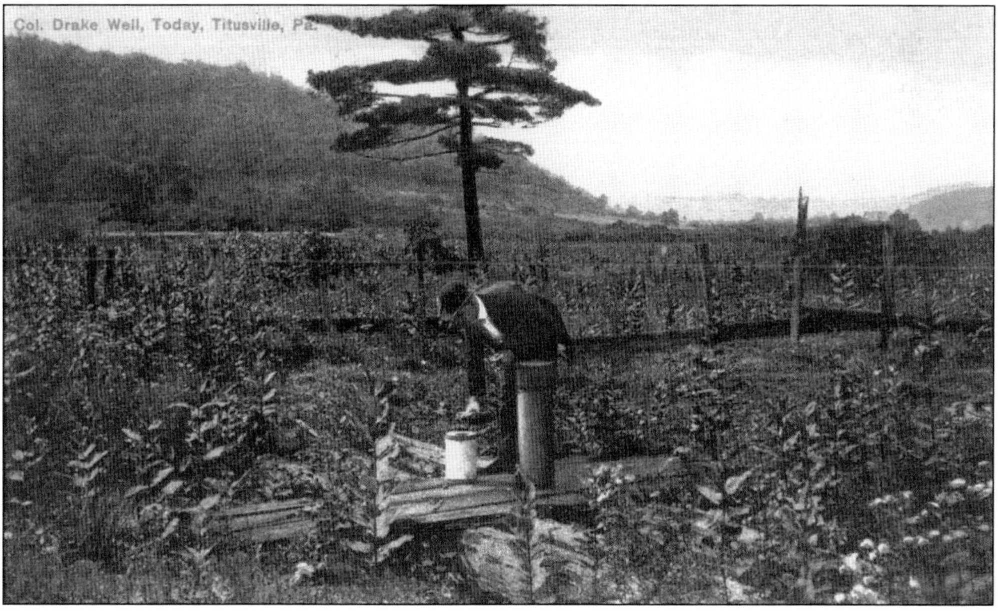

In the early 1900s, interest was growing in and around Titusville in commemorating both the contribution of Edwin L. Drake to the growth of the oil industry and the birth of the industry itself. At the time, little of anything remained to mark the historic site of Drake Well except for some casing and a few boards, as seen in this image. Indeed, Drake and his accomplishments seemed forgotten. This was about to change.

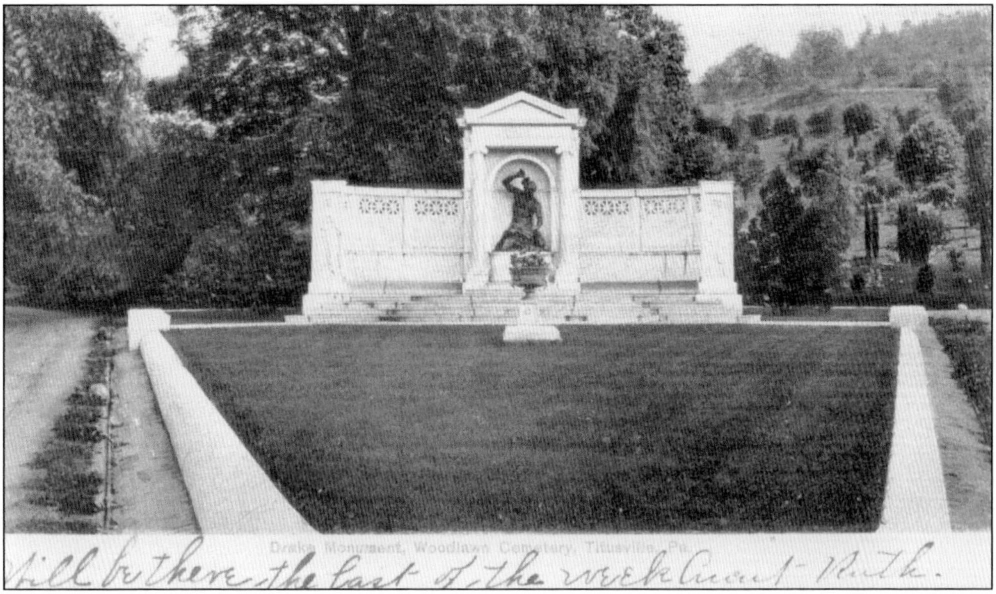

The Drake Monument, honoring Edwin L. Drake, was built at the Woodlawn Cemetery near Titusville in 1901. Henry H. Rogers of the Standard Oil Company donated $100,000 for its construction. The bronze statue framed by the cut stone columns is called *The Driller.* Drake died in 1880 and was buried in Bethlehem, Pennsylvania. He was moved back to Titusville in 1902 and reinterred. Drake and his wife, Laura, who died in 1916, lie in a vault before the monument.

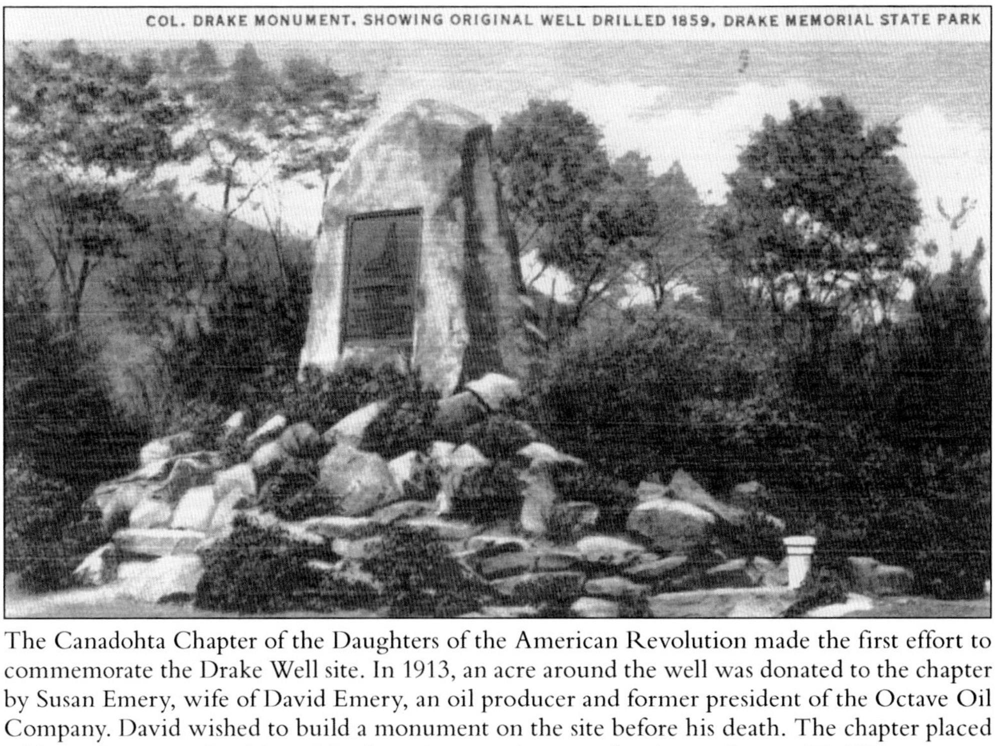

The Canadohta Chapter of the Daughters of the American Revolution made the first effort to commemorate the Drake Well site. In 1913, an acre around the well was donated to the chapter by Susan Emery, wife of David Emery, an oil producer and former president of the Octave Oil Company. David wished to build a monument on the site before his death. The chapter placed a 30-ton limestone boulder with a large bronze plaque at the site on August 27, 1914.

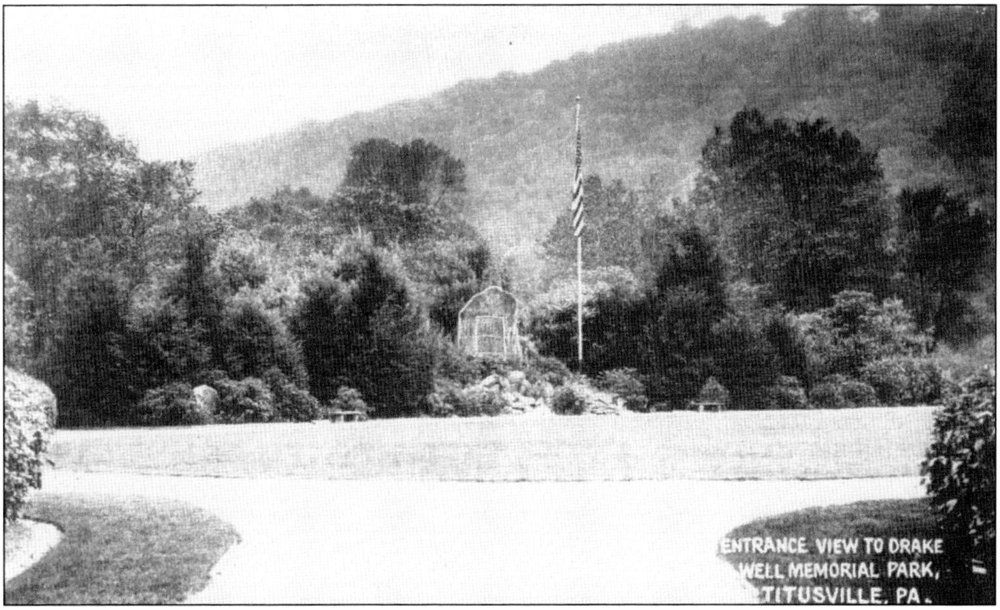

Local citizens maintained interest in building a museum and library at Drake Well to preserve the early history of the petroleum industry. In 1931, the American Petroleum Institute raised $60,000 to secure the site, establish a museum and library, and build a caretaker's house, with the goal of transferring the property to the state. In 1934, the property was deeded to the Commonwealth of Pennsylvania, ultimately to become Drake Well Park, administered by the Pennsylvania Historical and Museum Commission.

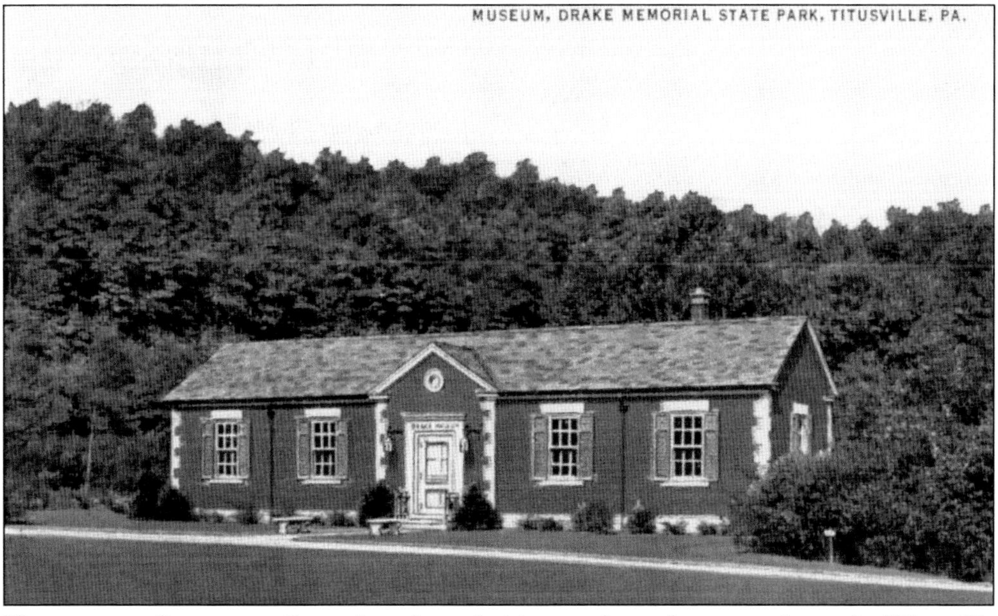

The concept of a museum focused on the early history of petroleum was the goal of Titusville resident E. C. Bell. Bell had collected petroleum memorabilia for over 50 years and had the best collection in the country. He built his own museum to house his materials near his home on Oil Creek. This postcard, postmarked 1939, shows the original museum at Drake Well Park, which housed tools, books, company records, and Mather negatives, among other items.

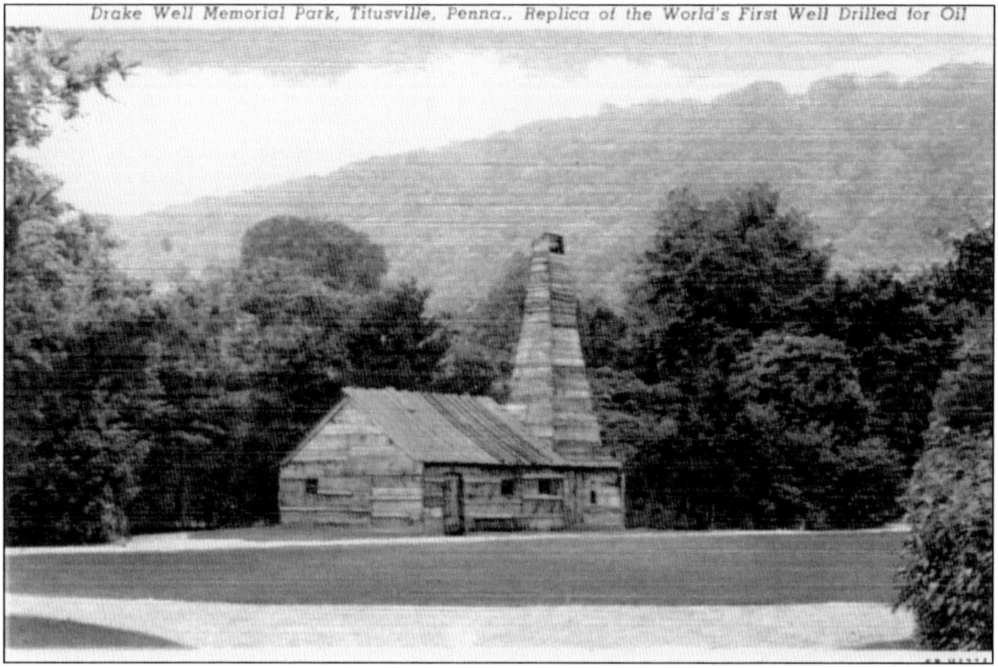

Visitors to Drake Well Park expressed interested in seeing what the well looked like in 1859. In 1945, a replica of Drake Well was built at the well site, including the derrick and engine house. In order to build the replica on the well's exact site, the stone monument donated by the Canadohta Chapter of the Daughters of the American Revolution was moved toward the entrance of the park. This 1930s postcard shows the replicated well.

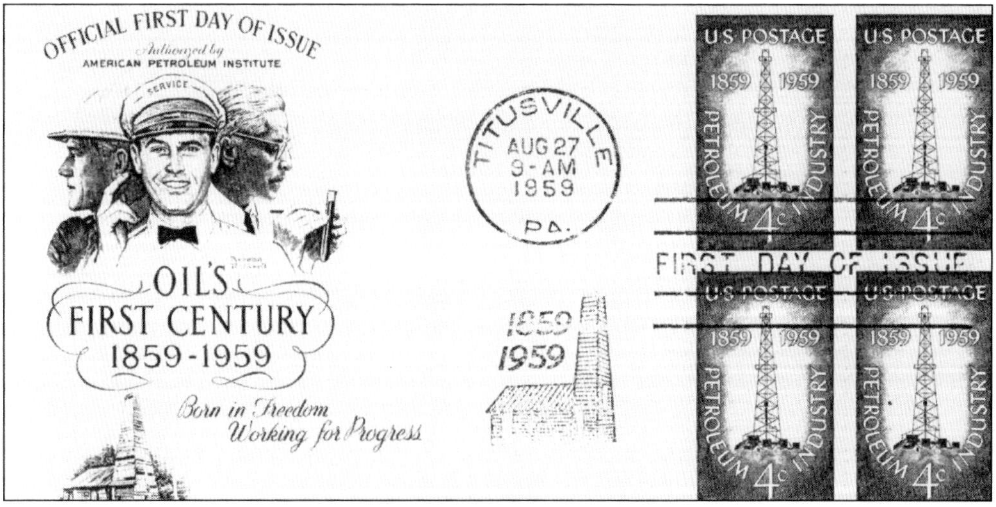

In 1959, the centennial of the birth of the oil industry was celebrated in Pennsylvania's oil region. To commemorate the event, the American Petroleum Institute authorized a first day of issue cover bearing a design by Norman Rockwell. The logo bears the slogan "Oil's First Century 1859-1959 Born in Freedom, Working for Progress." Today the Pennsylvania oil region is recognized and celebrated as both state and national heritage areas.

Six

The Oil City Region

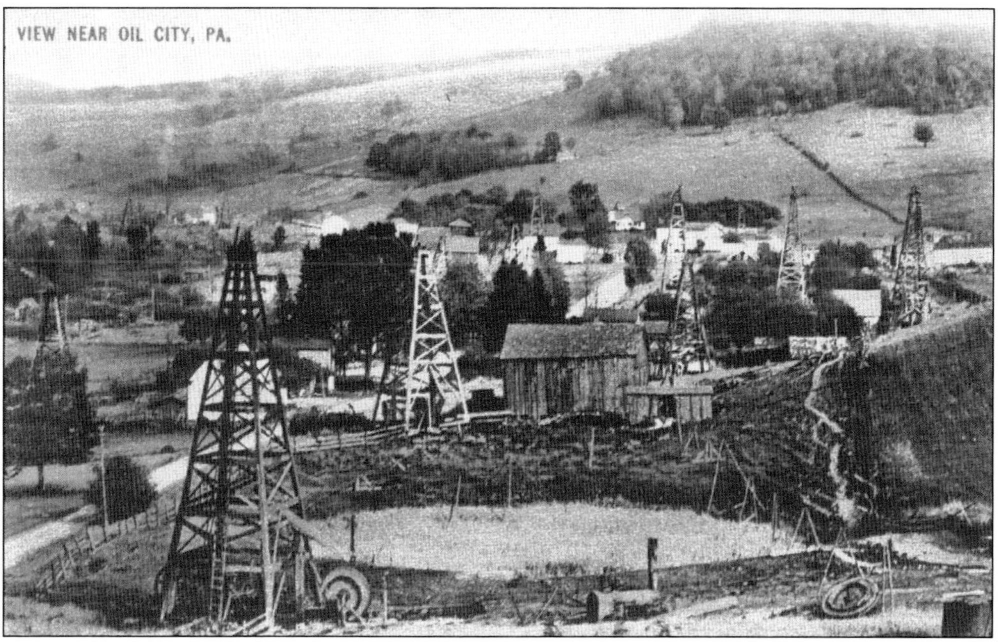

This quiet image of the countryside near Oil City understates the frenzy that hit the area during the oil boom in the Oil Creek Valley. During the boom, Oil City became an important staging area for the transport of oil, particularly by boat. As early as 1862, steamers, towboats, and hundreds of flatboats were carrying Oil Creek crude from Oil City to markets in Pittsburgh.

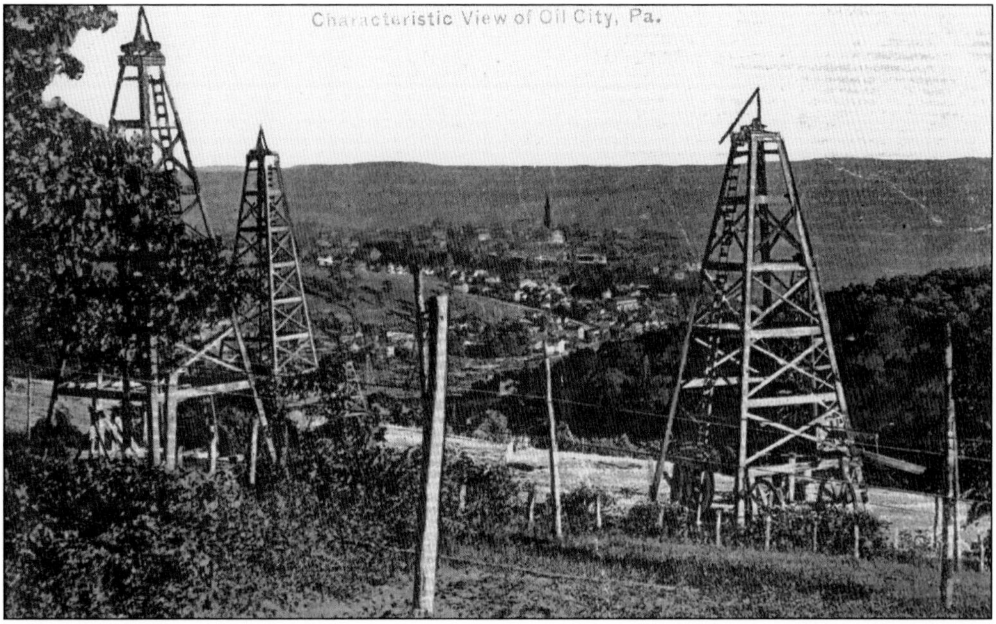

This early-1900s postcard shows a "Characteristic View of Oil City" with derricks in the foreground framing the twin steeples of St. Joseph's Church in the distance. J. R. G. Hazzard, a correspondent for the *New York Daily Tribune*, traveled around the Oil City area in 1868 noting, "The bottomlands are . . . a wilderness of derricks which remind you of . . . shipping crowded into some great harbor."

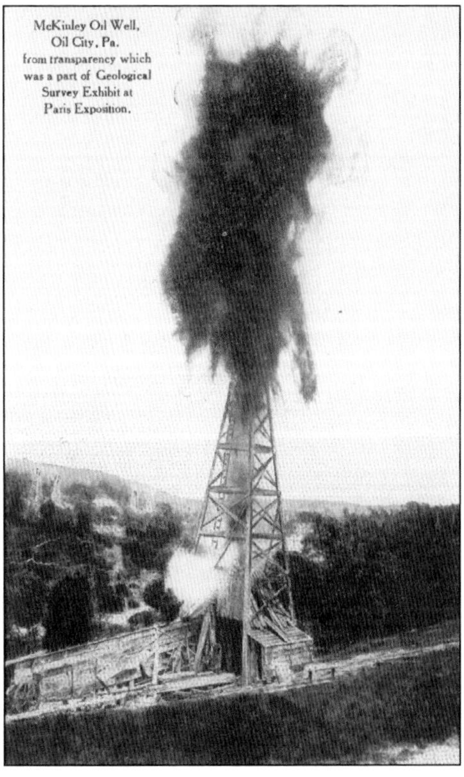

The McKinley Well in Oil City became world famous when a transparency of it flowing after being shot by a torpedo was shown as part of geological survey exhibit at the Paris Exposition of 1900. In 1865, the well was described as producing about 50 to 60 barrels daily with the stream fluctuating daily but never stopping altogether. This view of the McKinley Well was postmarked in 1913.

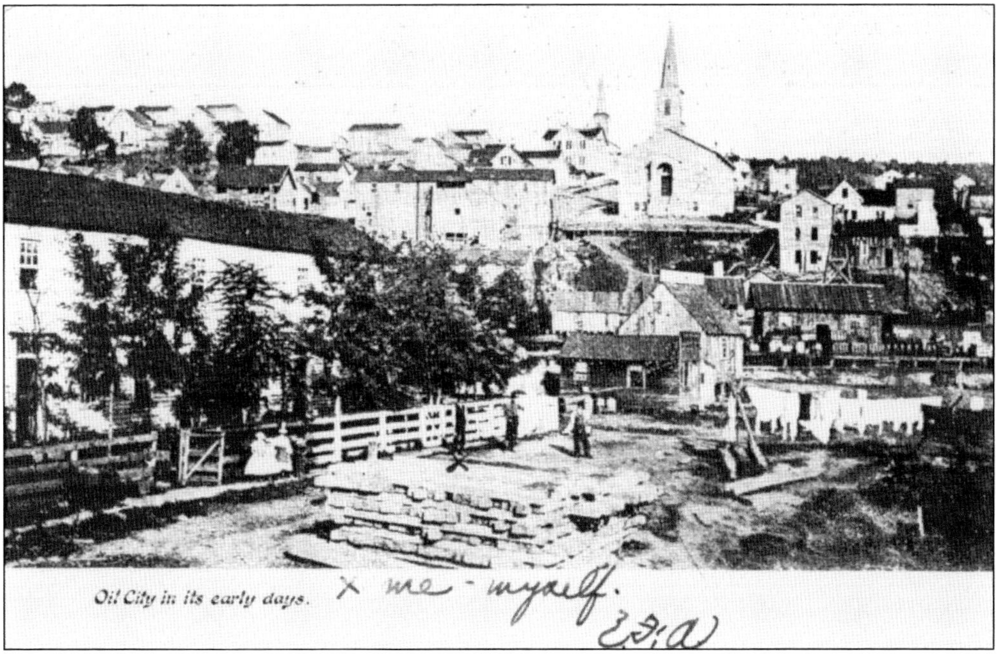

Located at the confluence of Oil Creek and the Allegheny River, Oil City developed into major shipping and processing centers for the region's oil. Cottage Hill, seen in this early image, was established on the east side of Oil Creek in 1863. The fervor for oil and profit at the time produced towns in which "the buildings were hurriedly constructed without any thought of architectural beauty or permanence." This postcard was postmarked in 1906.

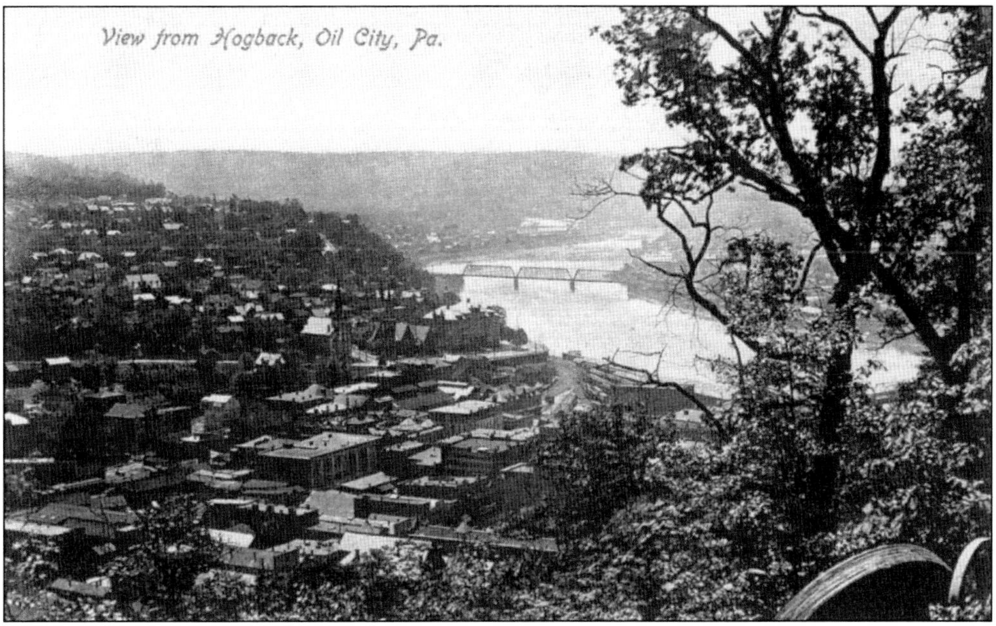

In the early 1900s, Oil City had a population of about 15,000. There were four refineries in Oil City at the time, including the Crystal, Continental, Independent, and Penn refineries. This view from "Hogback Hill," Oil City, postmarked in 1910, shows the city's North Side and Cottage Hill district.

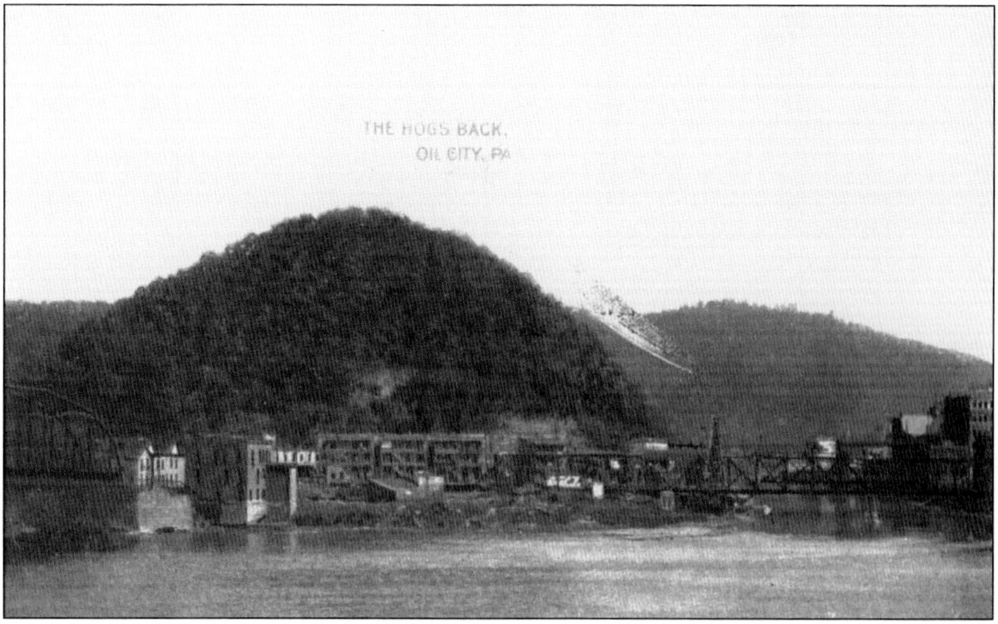

One of the prominent natural features in Oil City is the "Hog's Back" or Hogback Hill seen in the postcard above, which was postmarked in 1909. It flanks the western portion of the city's North Side. Another prominent feature of the North City, and vital to the growth of Oil City, was Oil Creek, seen in the bird's-eye view postcard below, postmarked in 1909. Note the derrick along the upper bank of Oil Creek.

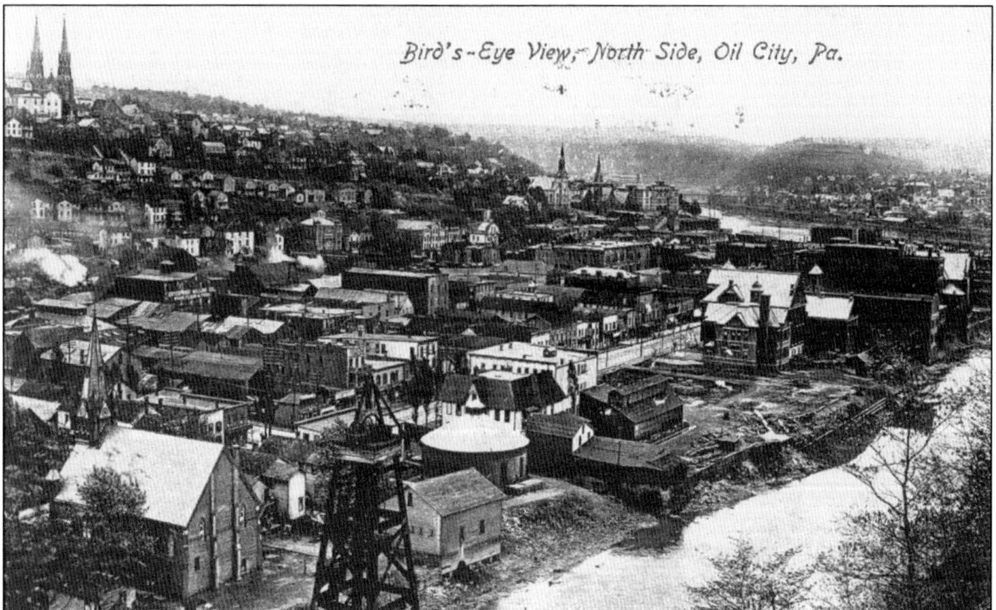

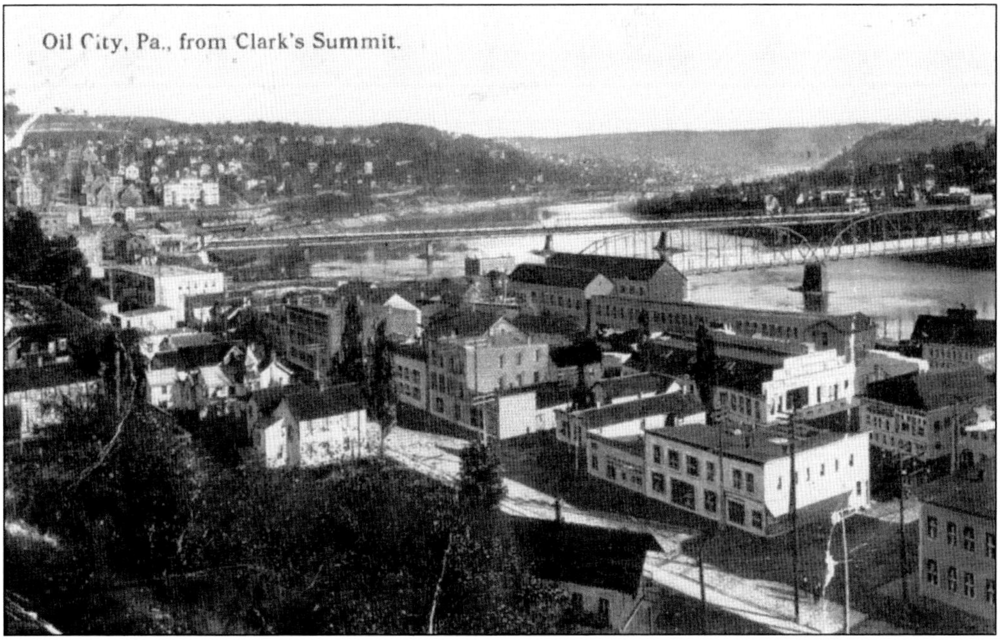

Oil City had variety of industries that supported its economy in the early 1900s. At the time there were barrel works, boiler shops, machinery shops, a foundry, woodworking shops, and three flour mills. The postcard above shows a view of Oil City from Clark's Summit in the early 1900s. The postcard below from the 1920s shows a bird's-eye view of Oil City's East Side.

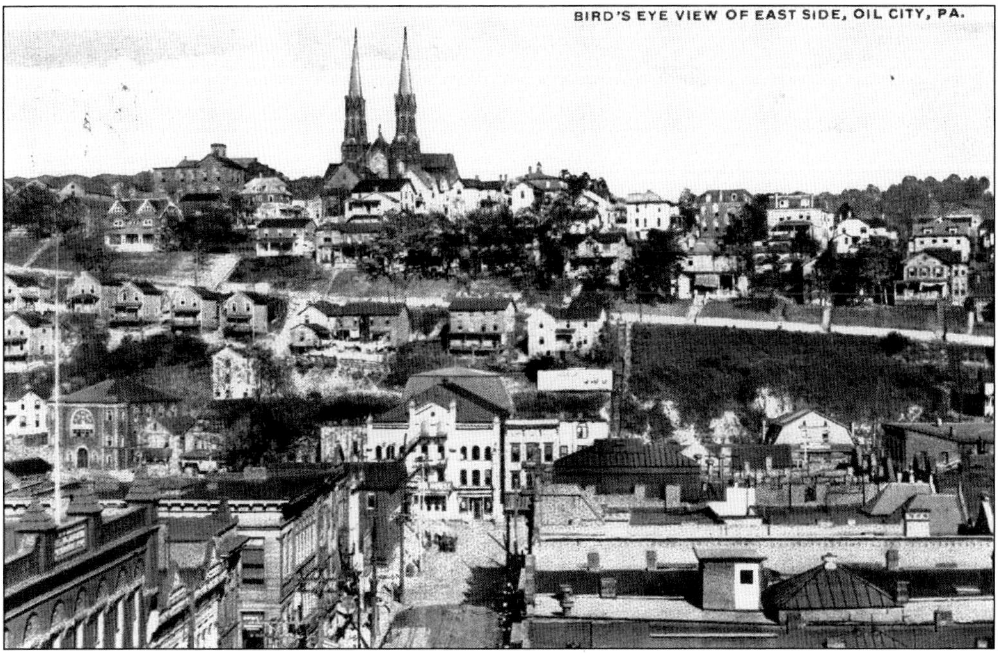

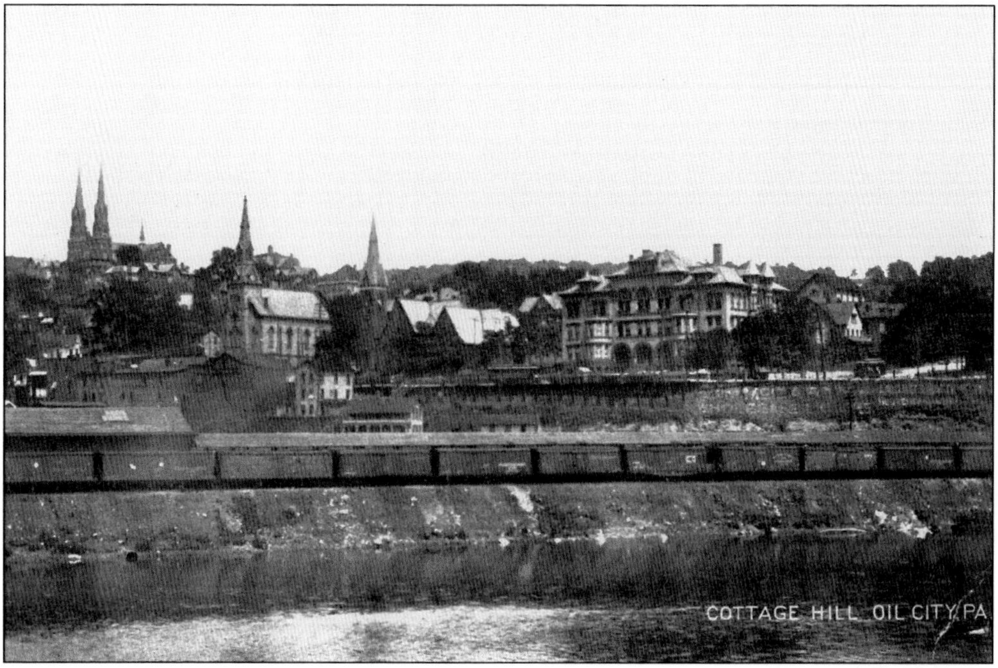

Another prominent landmark in Oil City's skyline is the twin steeples of St. Joseph's Church, seen in the distance in these two images. The church was built in 1890 and dedicated in 1894. The above postcard showing an image of Cottage Hill was postmarked in 1909. The image below of Center Street in Oil City is from the early 1900s.

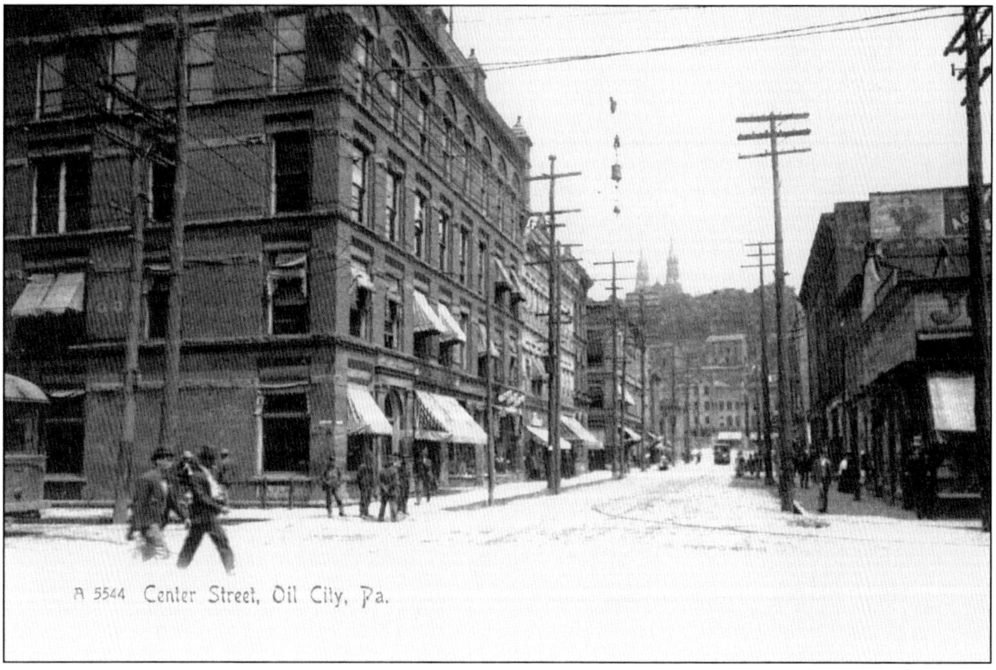

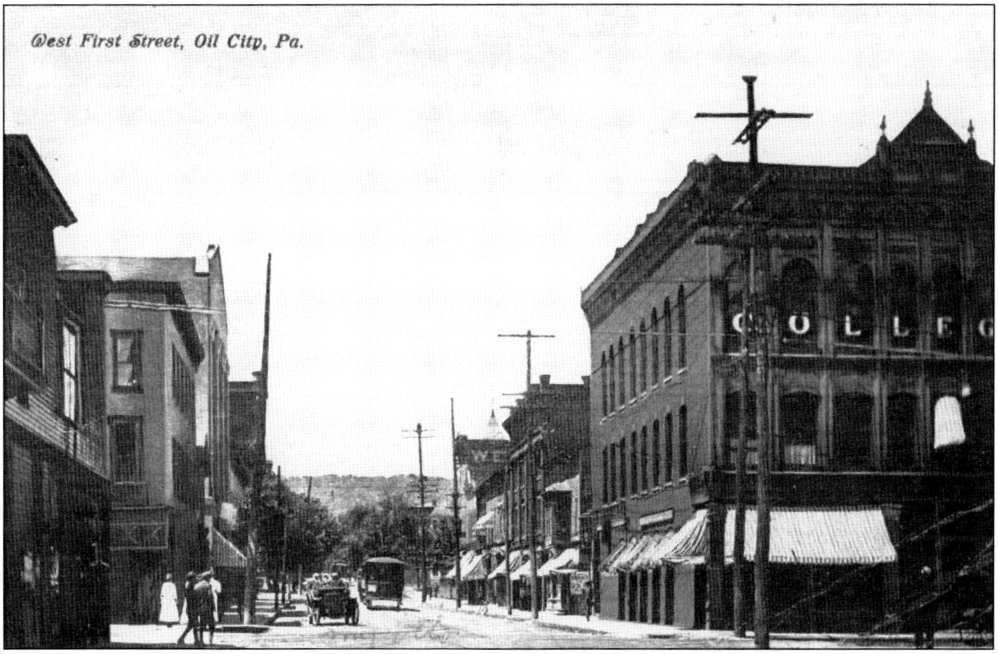

In 1864, a reporter traveling through the oil region had disparaging remarks for Oil City: "Whew! What smells so? . . . the gaseous wealth of the oily regions . . . I walked between walls and oil yards, barns and pens, keeping my bearing as I could . . . I found the main street . . . too thick for water and wholly too thin for land." By the early 1900s, Oil City had advanced far from its early days during the oil boom, as seen from these two images.

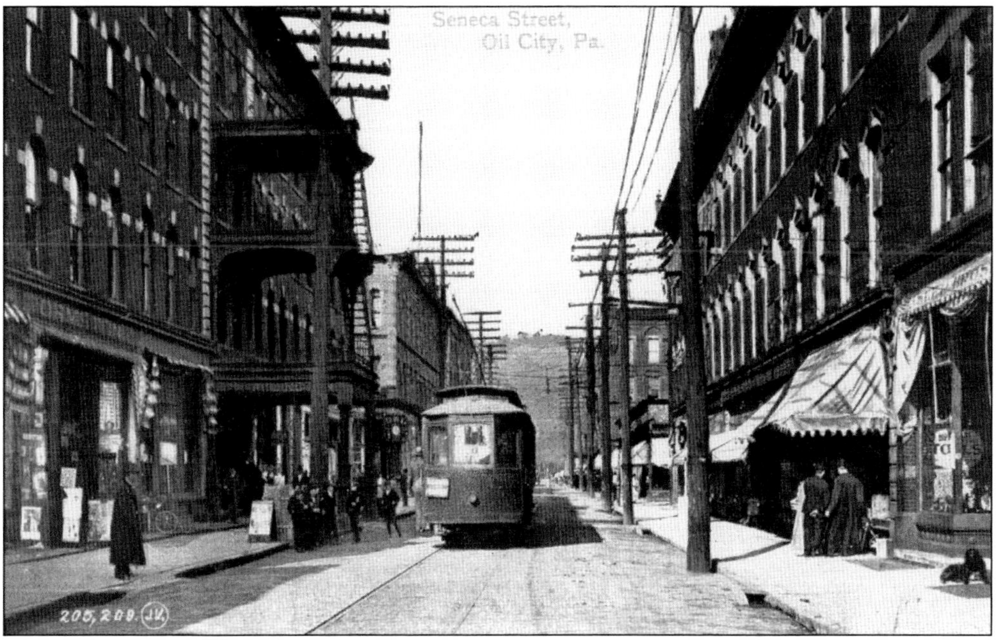

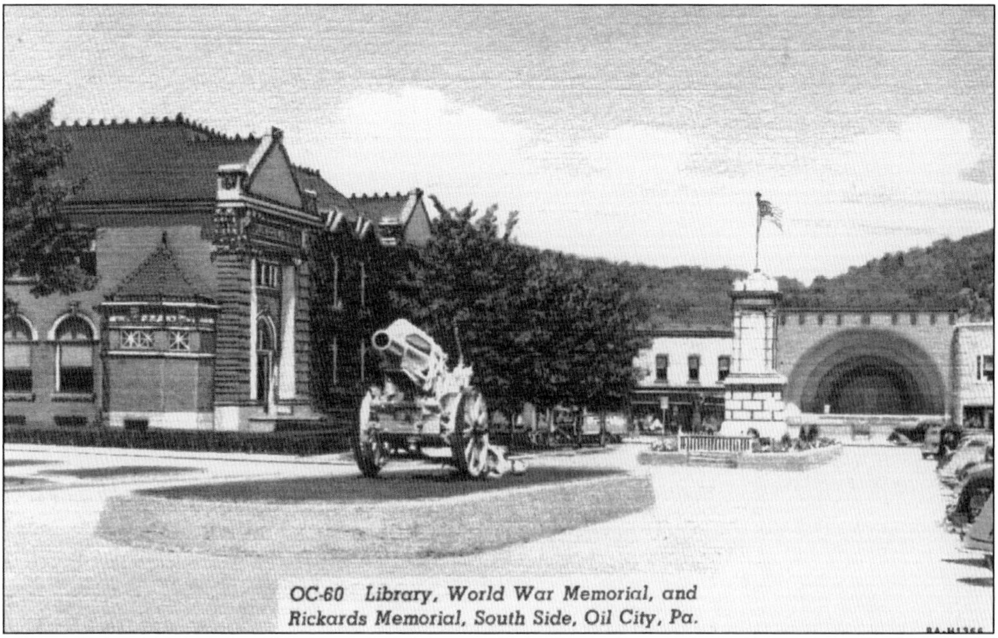

The Carnegie Library of Oil City, built in 1904, is located at the corner of Central Avenue and West Front Street. The library had its origins in 1890 with the Belles Lettres Club, a local literary group. A member of the club, Cora Hull Ramage, convinced Andrew Carnegie to donate $44,000 to build the library and $3,000 yearly for maintenance. This linen-era postcard of the Carnegie Library at Oil City also shows the World War Memorial and the Rickards Memorial.

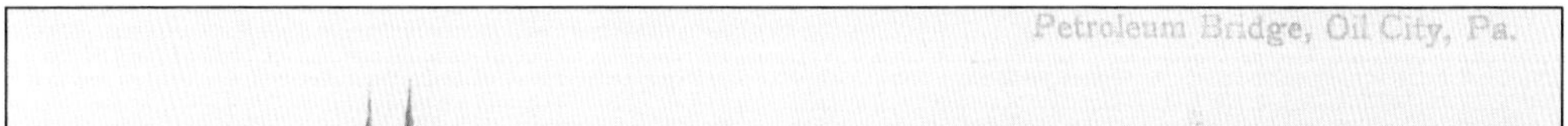

This postcard shows the Petroleum Street Bridge, which spans the Allegheny River just upstream from Oil Creek, connecting Oil City's South Side and North Side. The North Side of Oil City was one of the original three tracts of land given to Cornplanter, a Seneca chief in 1791. This tract was called the "Gift." Cornplanter sold this tract in 1818 for about $2,000. It is unclear whether Cornplanter ever received full payment for his land.

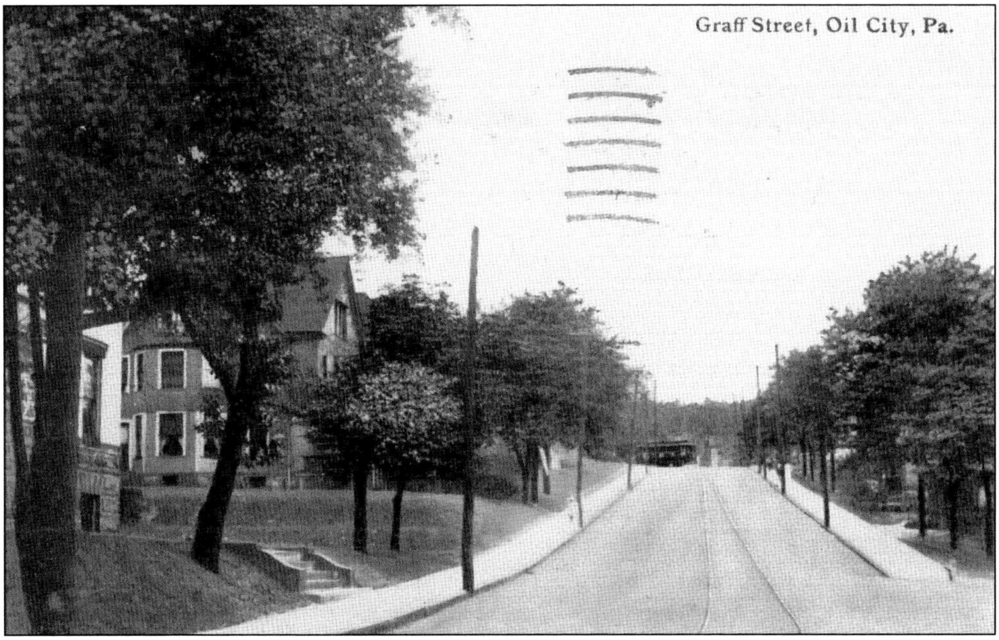

Graff Street is a residential street located in the Cottage Hill area of Oil City. The architecture in the area includes many late 19th- and 20th-century revival and late Victorian–style homes. The Cottage Hill area is within the Oil City North Side Historic District, established in 1999. Oil City also has a Downtown Historic District and a South Side Historic District. The view of Graff Street was postmarked in 1912.

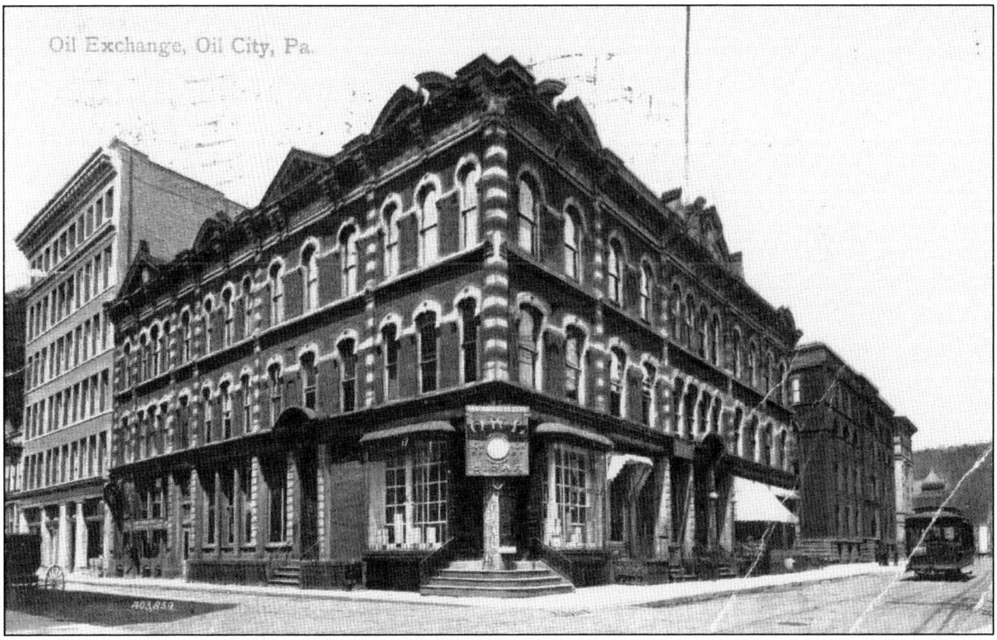

The Oil Exchange in Oil City was completed in 1878 and was located at the corner of Center and Seneca Streets. The building was lit by three large skylights and had a second-floor gallery for spectators that surrounded an elaborately frescoed exchange floor. The Oil City Oil Exchange was one of the last surviving exchanges in the oil region. It closed its doors permanently in 1908.

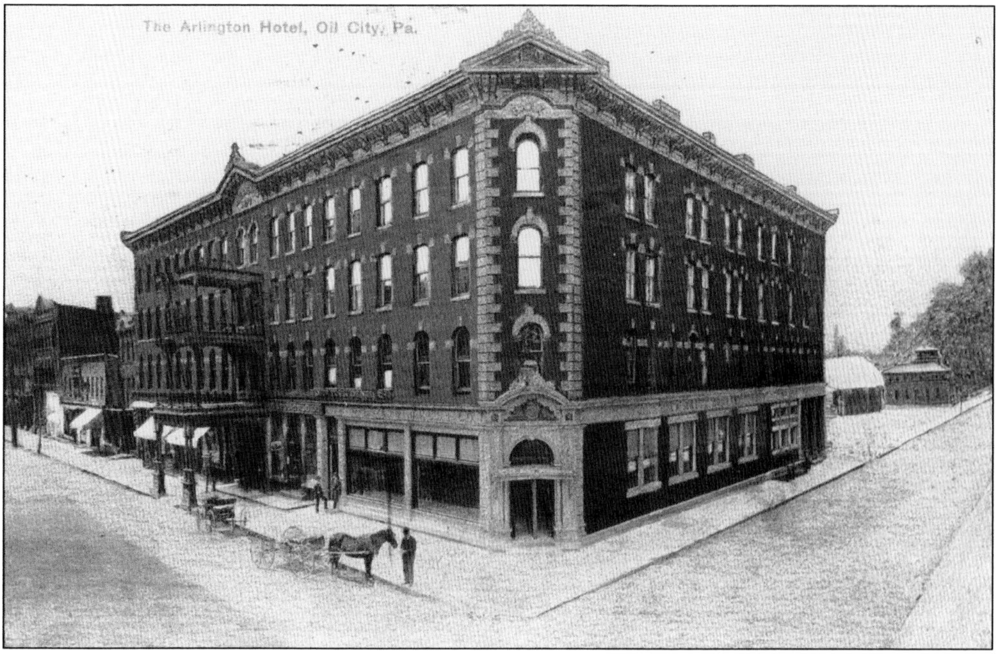

The Arlington Hotel in Oil City was one of the oil region's popular hotels. Michael Geary, a local industrialist, was the proprietor of the Arlington Hotel in the early 1900s. This image of the Arlington Hotel was postmarked in 1909.

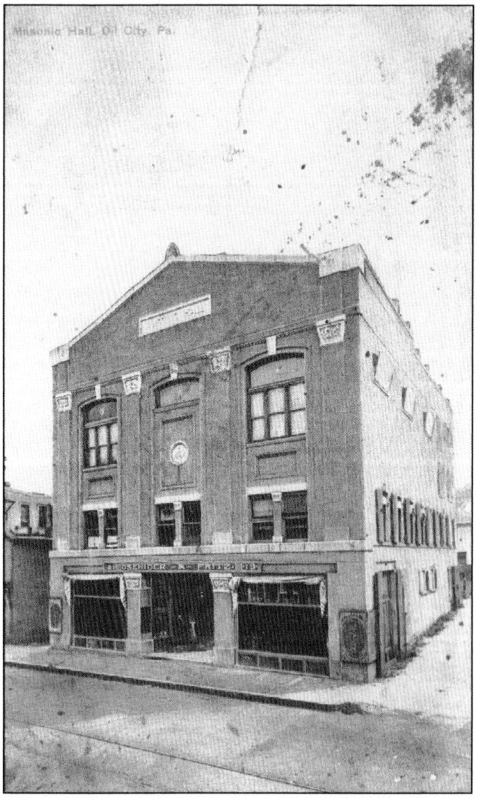

The Venango Perfection Lodge of the Scottish Right Masons in Oil City was constituted in 1904. Many of the leaders of the oil region have been members of this lodge. By 1910, the lodge had over 500 members. The Masonic hall, seen in this early-1900s image, is one of the classic buildings on First Street in Oil City's South Side.

The Oil City National Bank is a prominent building located on the corner of Seneca and Center Streets on the city's North Side. This postcard shows a view of the Oil City National Bank in the 1920s.

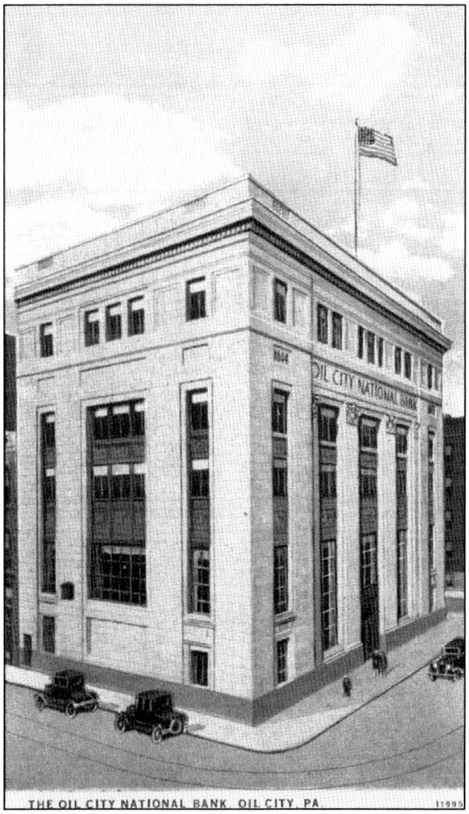

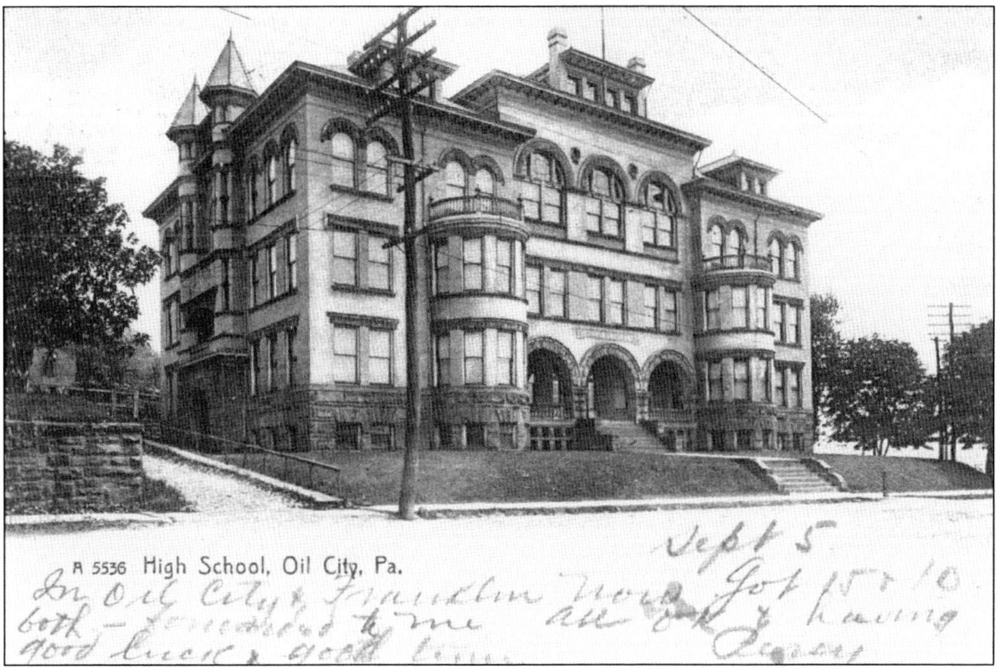

Oil City's old high school is located in the Cottage Hill area overlooking the Allegheny River. This postcard of the Oil City High School was postmarked in 1906.

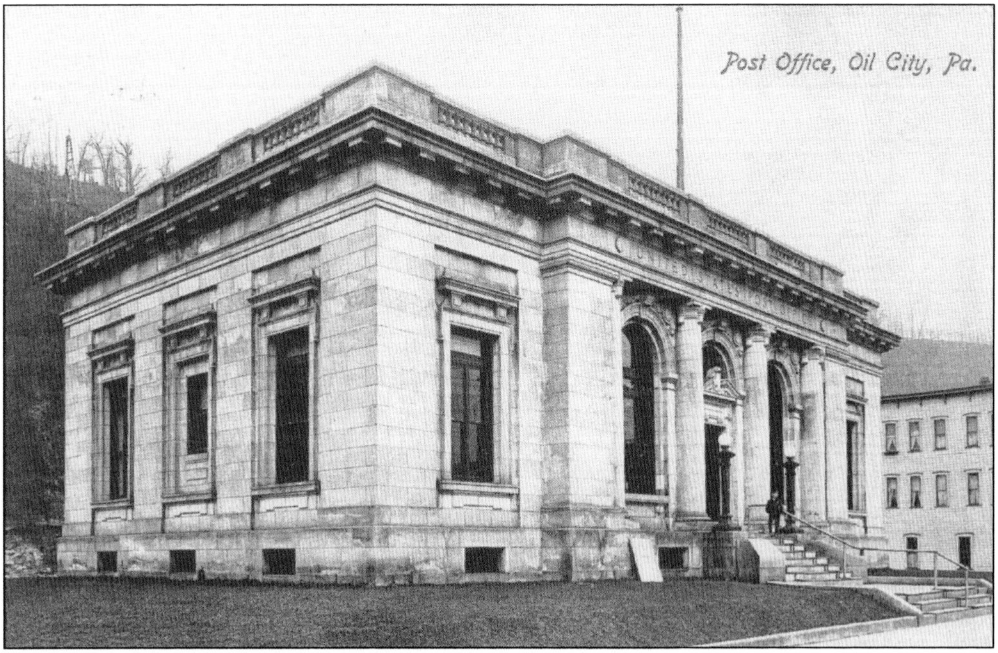

The former post office building, located on Seneca Street on Oil City's North Side, is an example of the Beaux-Arts style of architecture. It currently houses the Venango Museum of Art, Science and Industry. This image of the Oil City Post Office was postmarked in 1909.

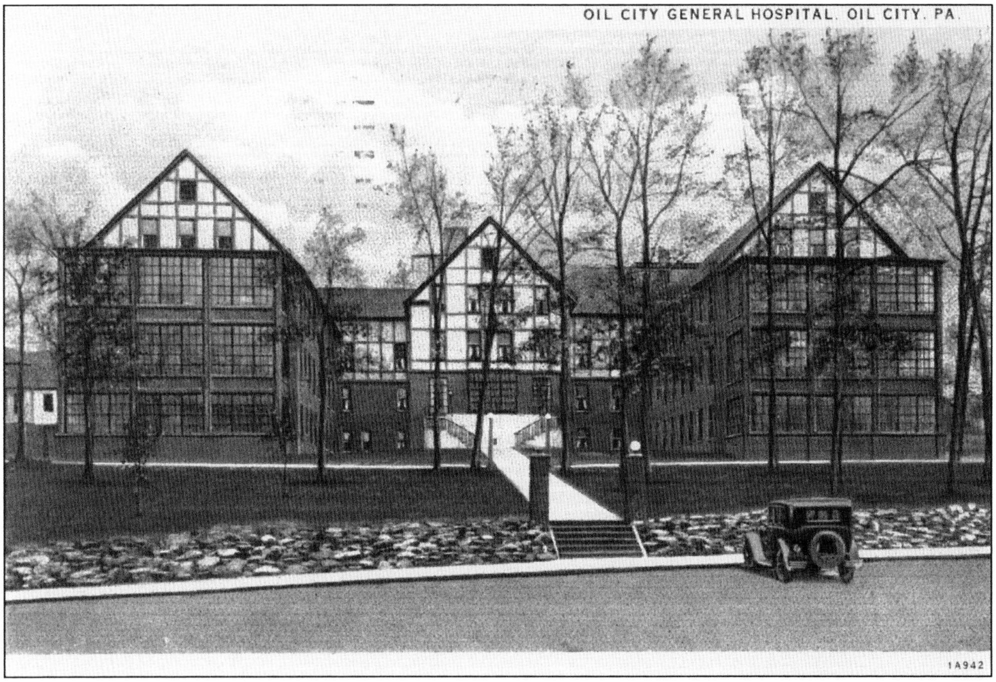

Henry H. Rogers started in the oil business with a small refinery at Rouseville. He ultimately rose to prominence in the Standard Oil Company. Rogers helped establish the Oil City General Hospital and a training school for nurses. This postcard shows the hospital in 1933.

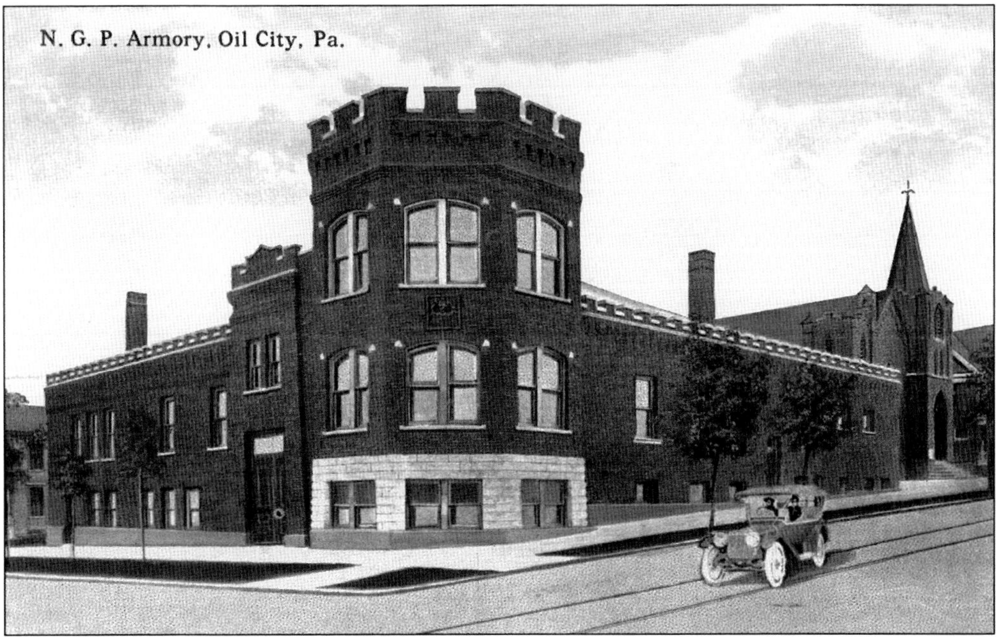

The Oil City Armory was built in 1900 and is a fine example of the late Victorian architecture that is a hallmark of much of Oil City. The armory was designed by architect Emmett E. Bailey, who designed other public buildings in western Pennsylvania.

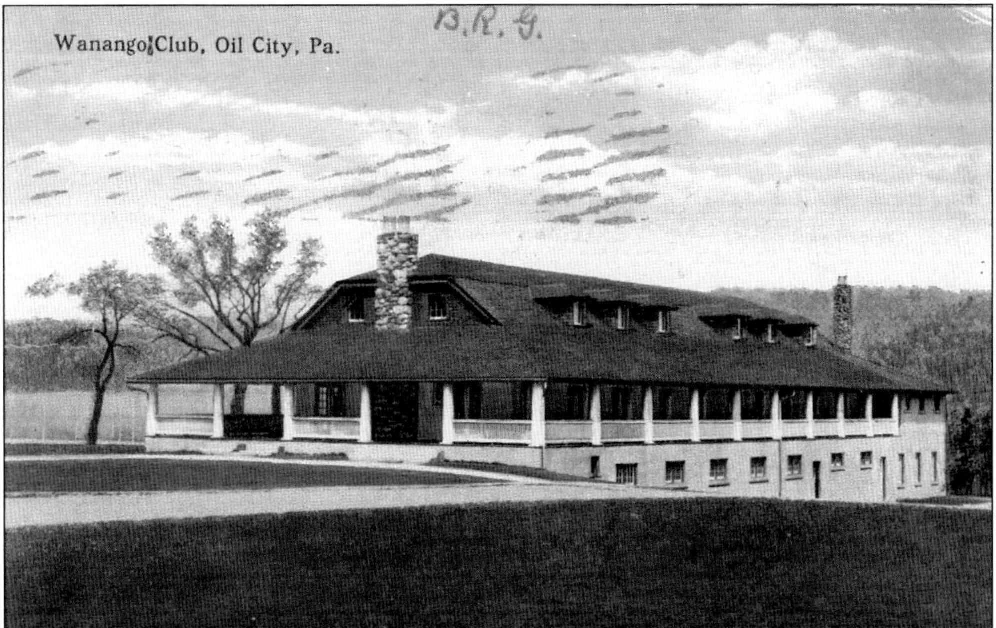

The Wanango Country Club near Oil City was started in 1912 by community leaders from Oil City and Franklin. Donald J. Ross was hired to design the first nine holes of the course. Ross had designed the famous Pinehurst golf course in North Carolina. This postcard, postmarked 1912, shows the Wanango Club's clubhouse.

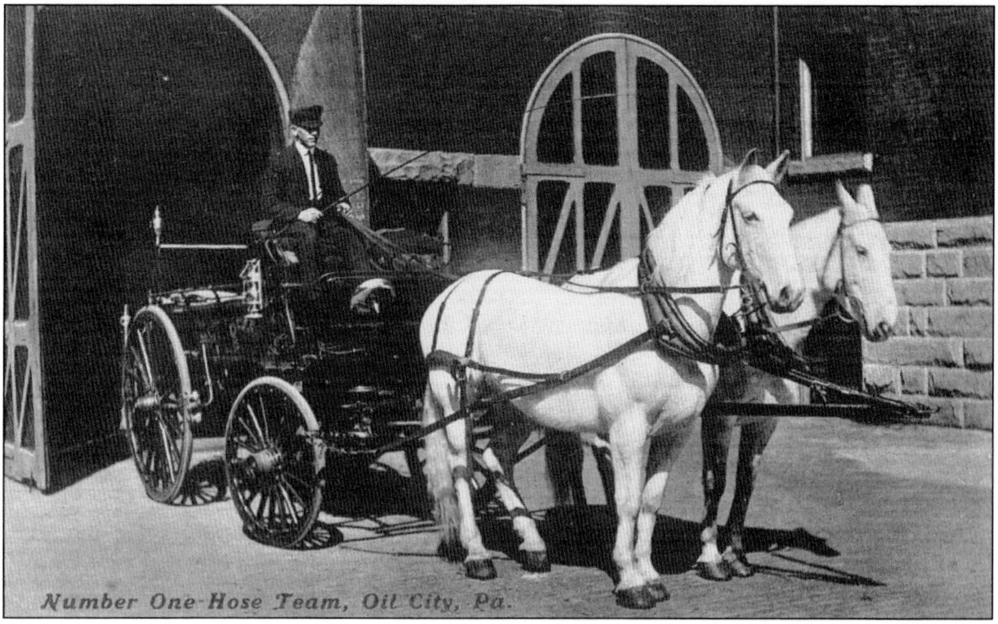

Oil City was hit hard by fires several times in its early days. Among the worst was the fire and flood of 1892 that devastated Oil City and Titusville. In Oil City, 50 people lost their lives, over 800 people became homeless, and property damage was extensive. This postcard shows an image of the "Number One Hose Team" of Oil City in the early 1900s, part of the vanguard that fought fires in the oil region.

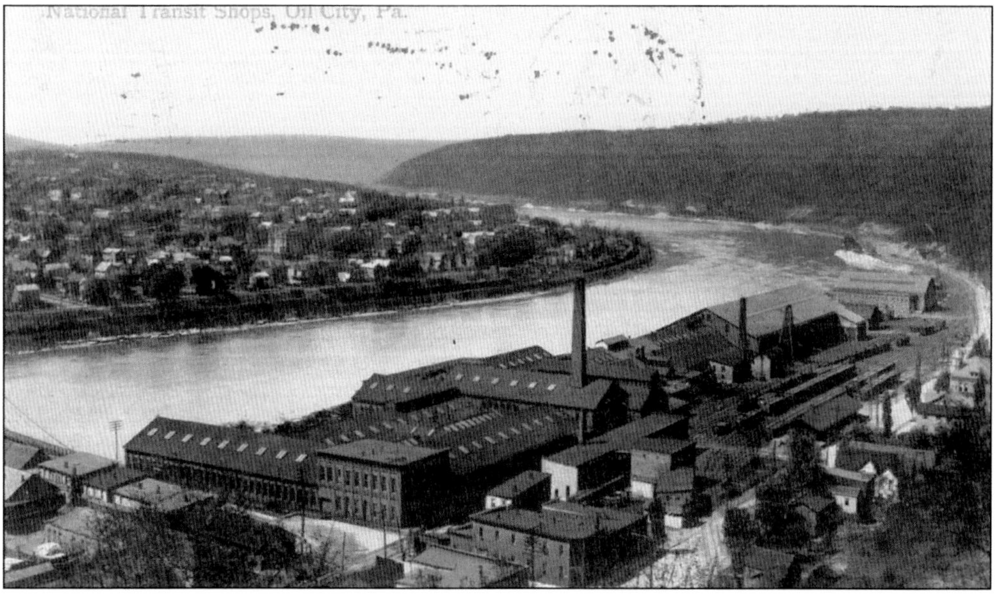

The National Transit Company was an affiliate of Standard Oil Company, for which it supplied parts and made repairs until the breakup of the trust in 1911. It then became the National Transit Pump and Machine Company. National Transit Pump and Machine Company made compressors for the region's gas industry, some of the largest reaching 1,000 horsepower. Compressors create the force needed to pump gas through pipelines. This image of the National Transit Shops was postmarked in 1910.

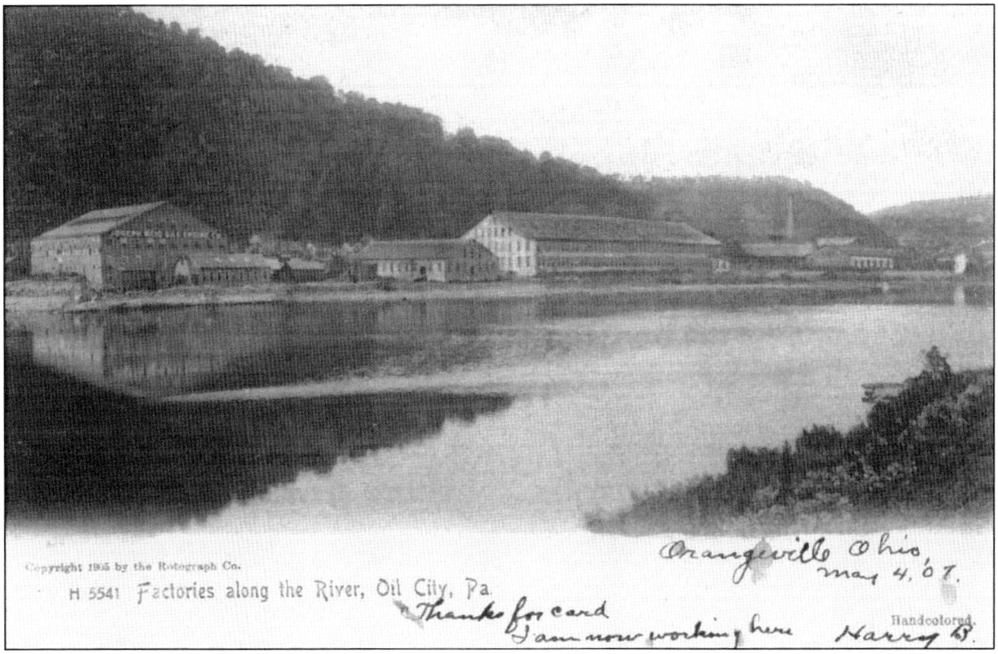

This image of Oil City, "Factories along the River," was postmarked in 1907 and shows the Joseph Reid Gas Engine Company on the left and the National Transit Pump and Machine Company shops on the right. The Joseph Reid Gas Engine Company was incorporated in 1898 and built engines for use in the oil fields. Joseph Reid immigrated to the United States in 1862 from Scotland. He was highly successful in designing quality oil field machinery and held numerous patents.

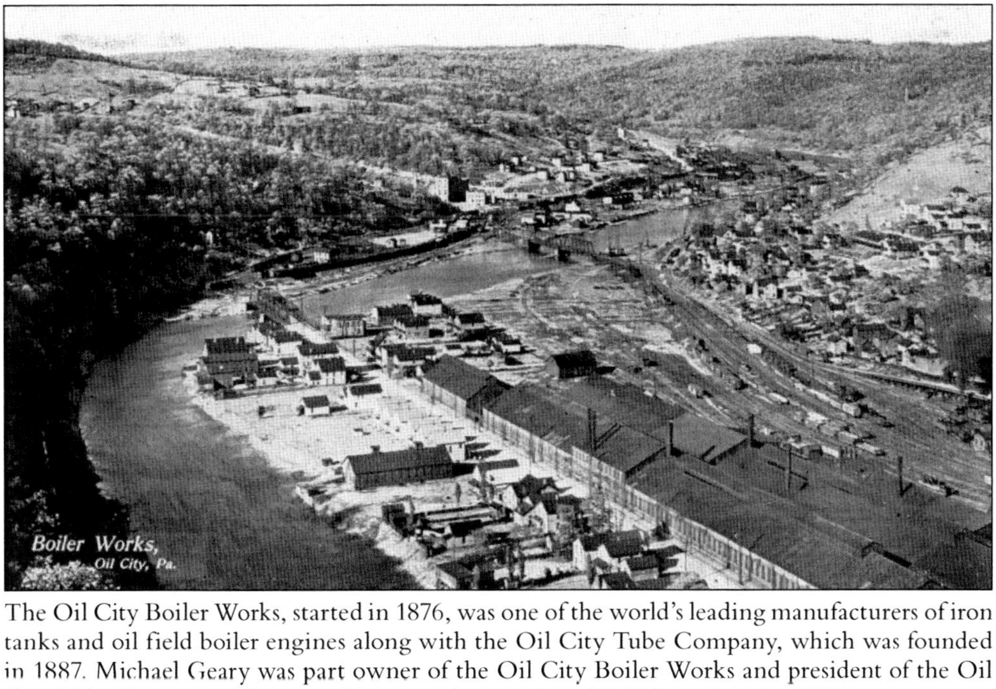

The Oil City Boiler Works, started in 1876, was one of the world's leading manufacturers of iron tanks and oil field boiler engines along with the Oil City Tube Company, which was founded in 1887. Michael Geary was part owner of the Oil City Boiler Works and president of the Oil City Tube Company. The two plants together employed 2,000 workers in 1895.

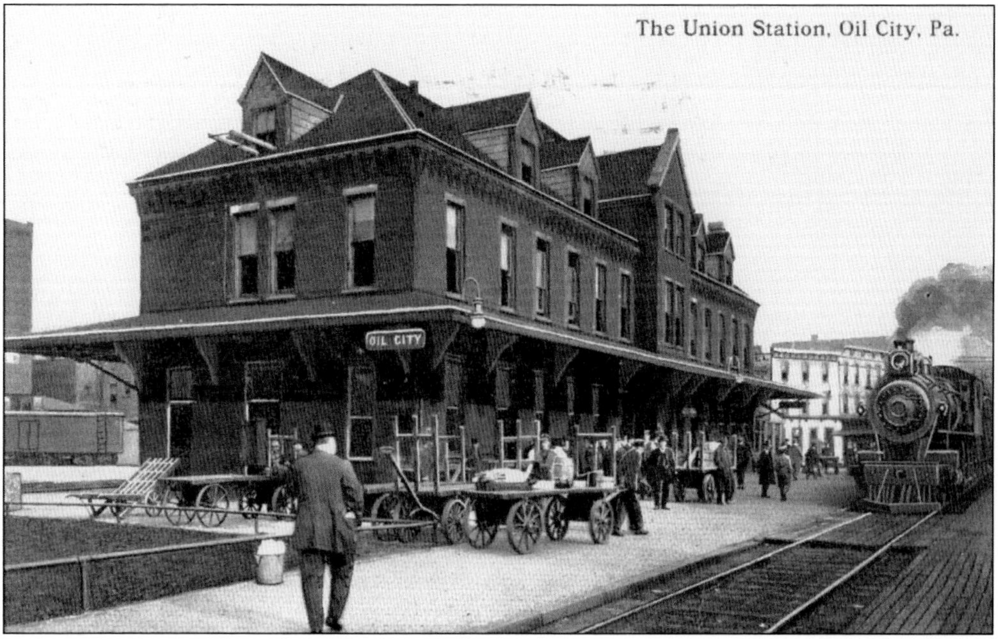

The hub of the oil region, Oil City was well served by rail lines in the early 1900s. Several rail lines intersected in Oil City, including the Erie, the Lake Shore and Michigan Southern, and the Allegheny Division of the Pennsylvania Railroad. The union depot in Oil City, which served the Pennsylvania Railroad, is shown in the postcard above, postmarked in 1914. The postcard below shows the union depot and nearby Sycamore Street.

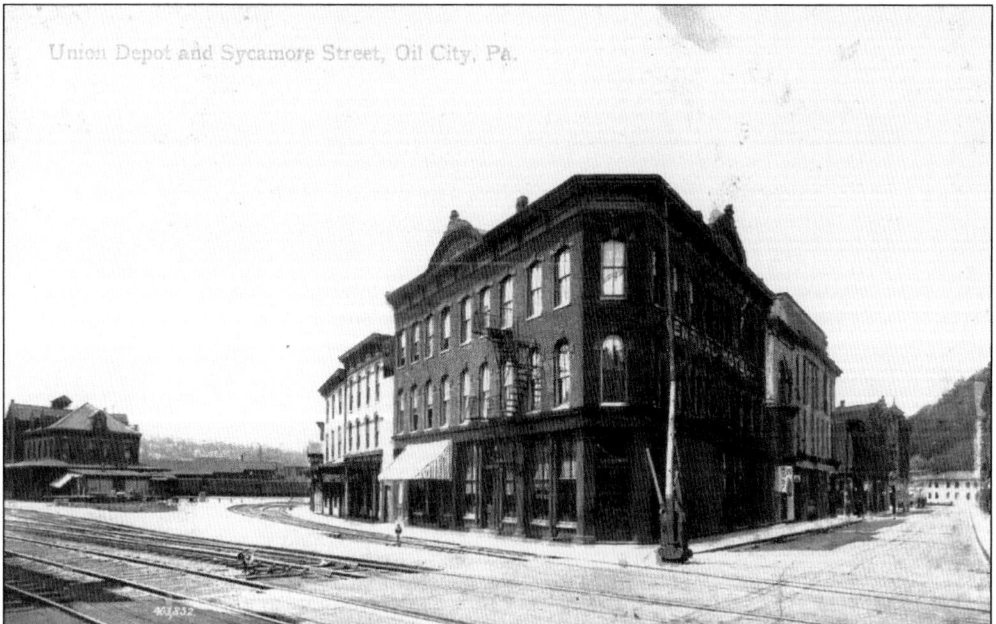

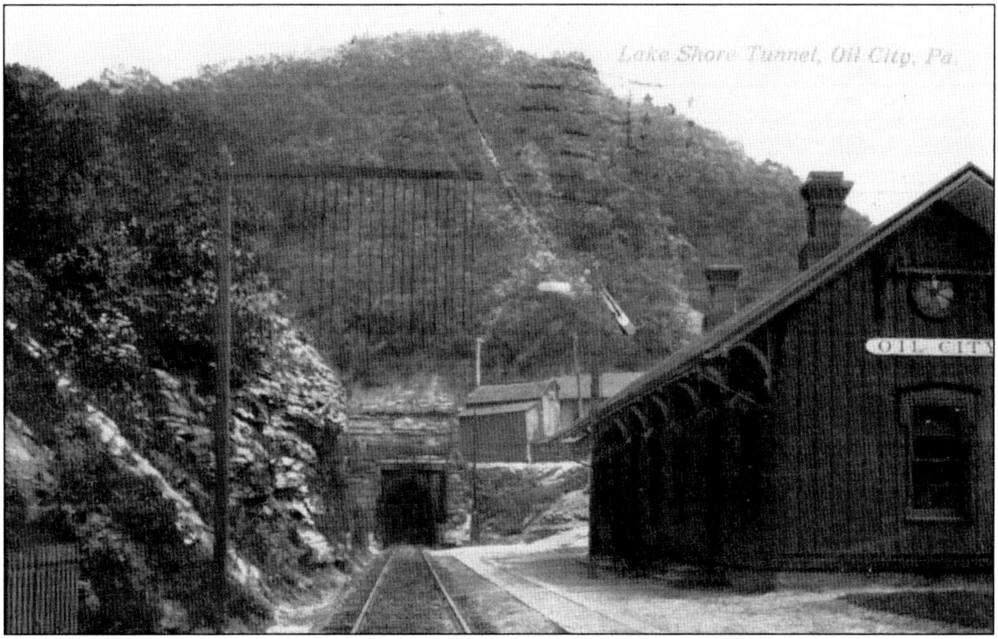

During the 1860s, the Lake Shore and Michigan Southern Railway extended a branch from Ashtabula, Ohio, to Oil City and Franklin. John D. Rockefeller shipped oil from Pennsylvania to his refineries in Cleveland on this line. The railroad used a tunnel built in 1870 to pass through Oil City's Hogback Hill. The postcard above, postmarked 1913, shows the railroad's depot and entrance to the tunnel. The postcard below, postmarked 1912, shows a train exiting the tunnel along Oil Creek.

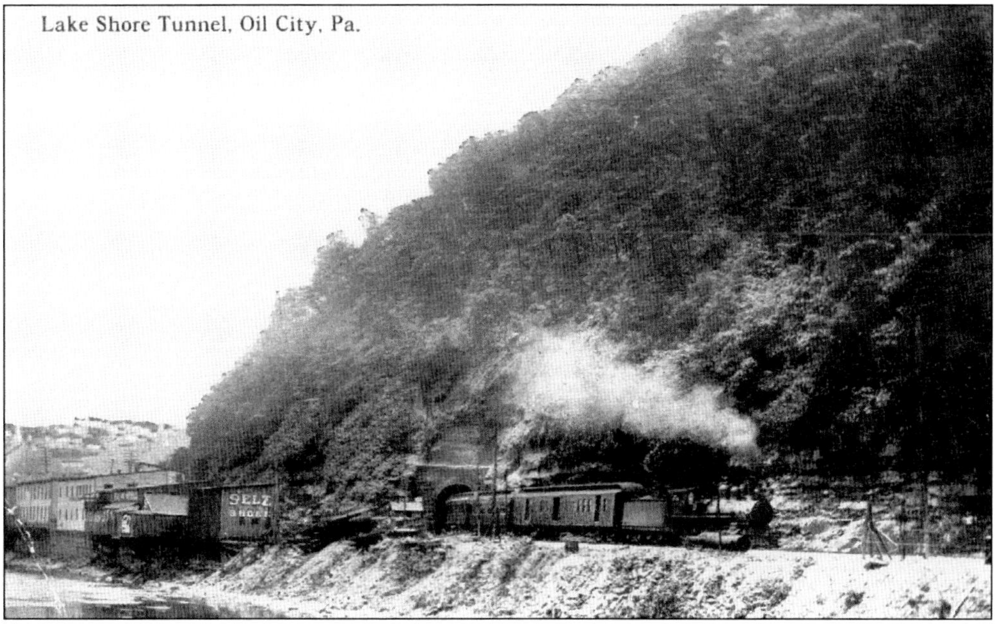

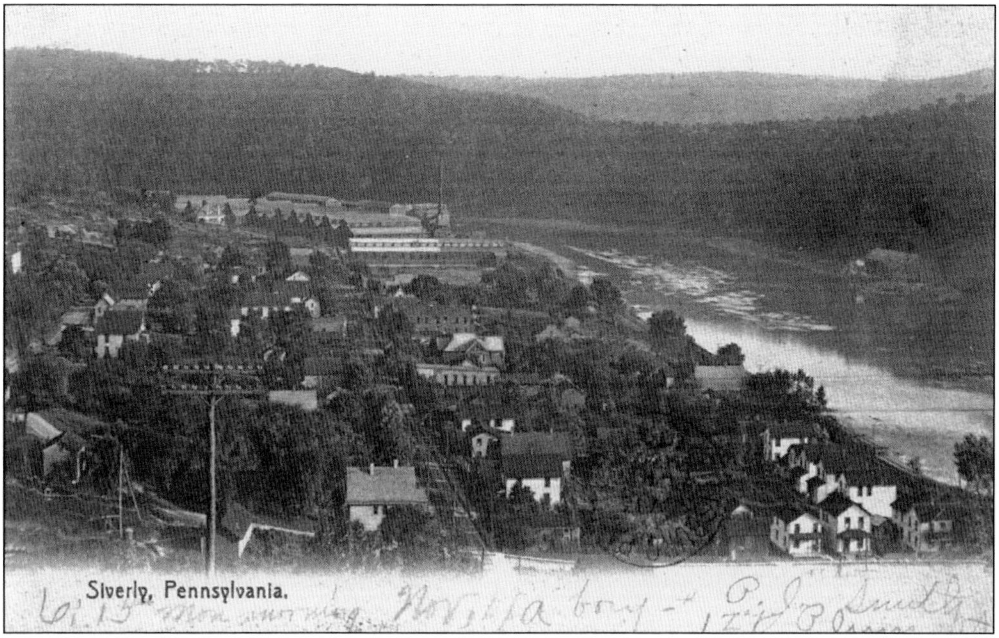

Siverly was a borough on the west side of the Allegheny River adjoining Oil City. It had a population of about 1,400 in the early 1900s and consisted of a residential section on the bluff above the river and a business section next to the river. The major industry at the time was the Oil Well Supply Company, which manufactured oil and gas well tools and machinery. It employed 650 workers.

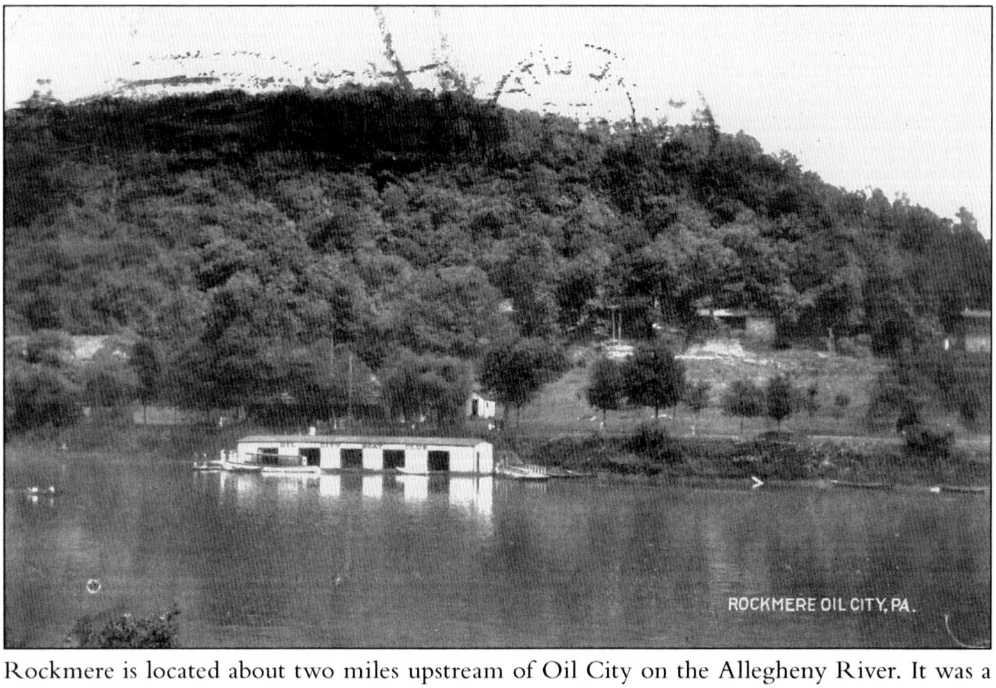

Rockmere is located about two miles upstream of Oil City on the Allegheny River. It was a popular area for summer homes and recreation, especially boating. This view of the Allegheny River at Rockmere, postmarked 1909, shows the boathouse of the Oil City Boat Club.

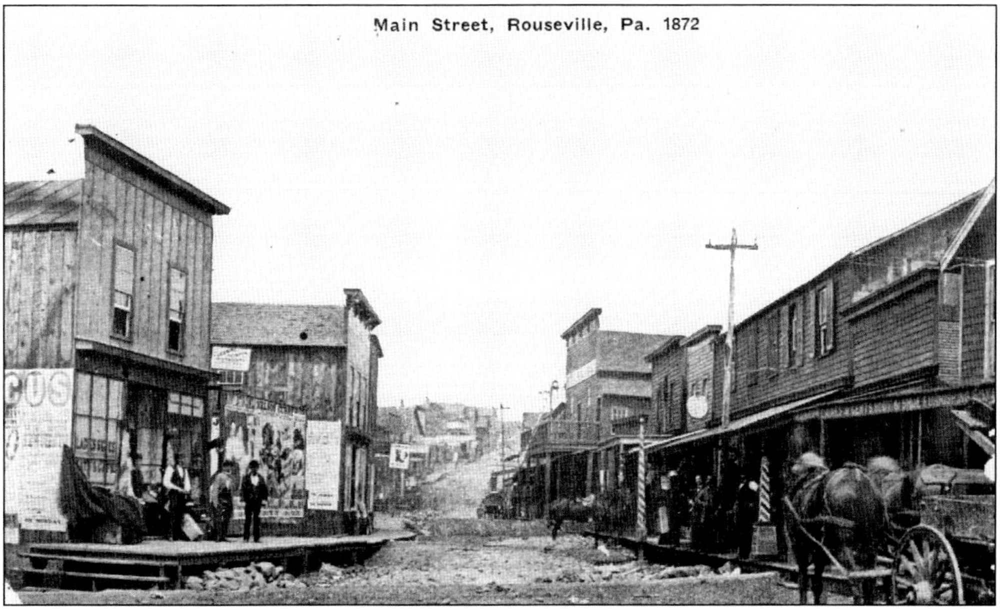

Rouseville, north of Oil City on Oil Creek, was named after Henry R. Rouse. Rouse was Warren County schoolteacher, lumberman, and legislator, who sunk the third well in Pennsylvania on the Buchanan Farm in 1859. Rouse was killed in a well fire in 1861, leaving much of his wealth in trust to Warren County. This postcard shows an image of Main Street, Rouseville, in 1872.

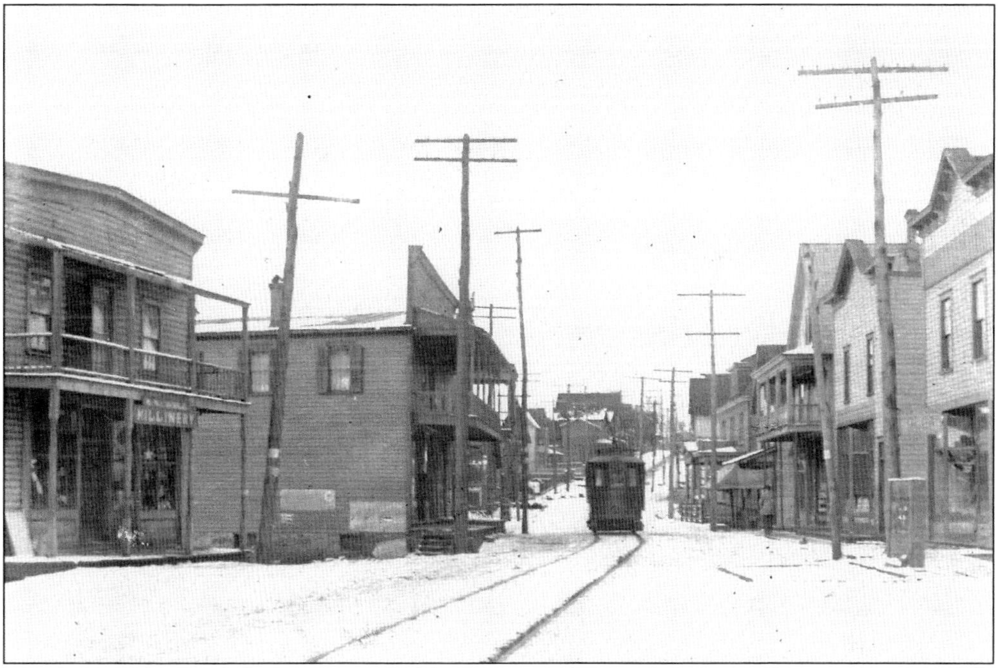

In 1915, Rouseville had a population of 700. There were two refineries in town at the time: the Germania Refining Company with 50 employees and the Crystal Oil Works with 18 employees. Germania Refining merged with Penn Refining in 1914, keeping the Germania name. It changed its name in 1917 after the United States entered World War I against Germany to avoid the bias of the time against anything "German."

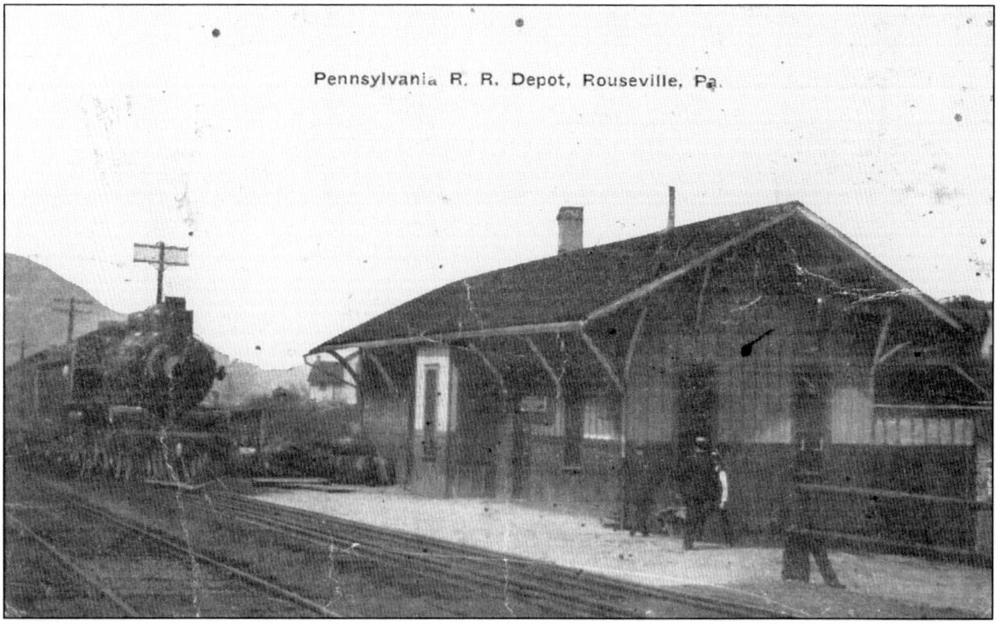

Rouseville was an important stop on the Oil Creek Railroad when the line was extended from Titusville to Oil City. The Pennsylvania Railroad also reached Rouseville in the late 1800s. The Pennsylvania Railroad depot at Rouseville is seen in this early-1900s postcard.

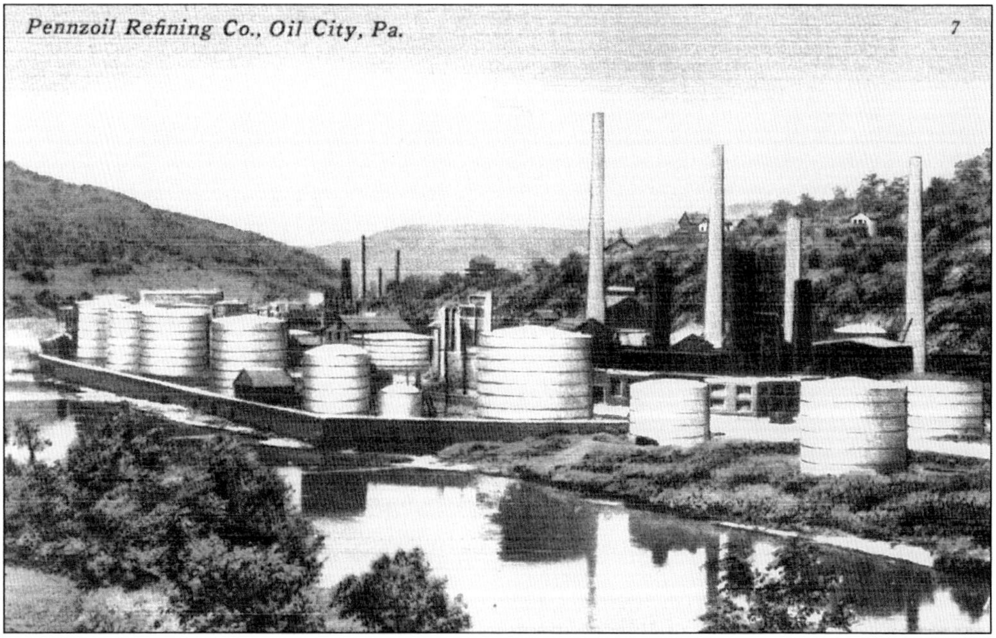

Rouseville was synonymous with the Pennzoil Company since the 1920s. The Pennzoil Refinery at Rouseville, depicted in the linen card above, was among the largest in Pennsylvania in 1940. It made a variety of products from crude oil ranging from wax to gasoline. It used the process of fractionization, or multiple distillation, by heat, steam, or vacuum, in the refining of crude.

Seven

Lower Allegheny River Oil Towns

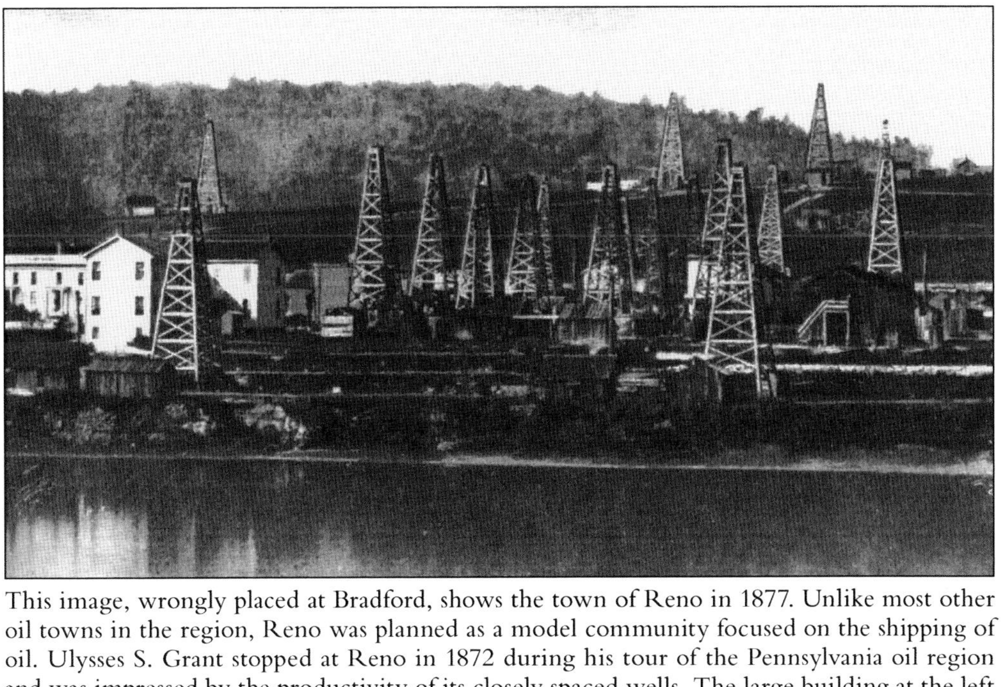

This image, wrongly placed at Bradford, shows the town of Reno in 1877. Unlike most other oil towns in the region, Reno was planned as a model community focused on the shipping of oil. Ulysses S. Grant stopped at Reno in 1872 during his tour of the Pennsylvania oil region and was impressed by the productivity of its closely spaced wells. The large building at the left of this image is the Folger House, a popular hotel of the time.

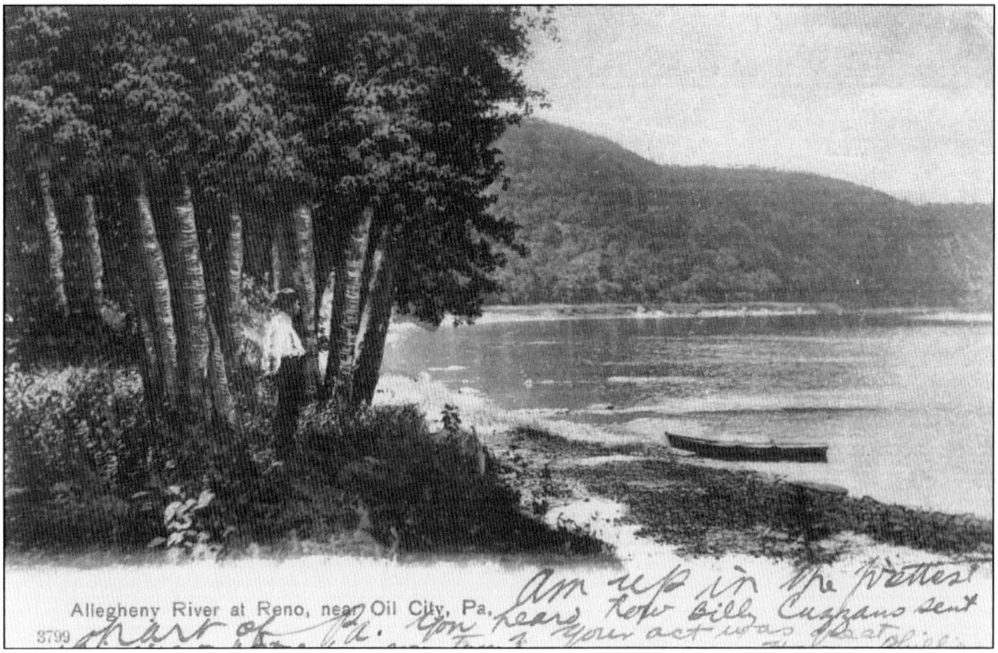

Reno is located between Oil City and Franklin on the Allegheny River. Banker and promoter Charles Vernon Culver envisioned Reno as an oil production boomtown, linked to the rest of the region by the Reno, Oil Creek and Pithole Railroad. The railroad was never completed, but Reno did become an important refinery town. This postcard, postmarked 1907, shows a view of the Allegheny River at Reno.

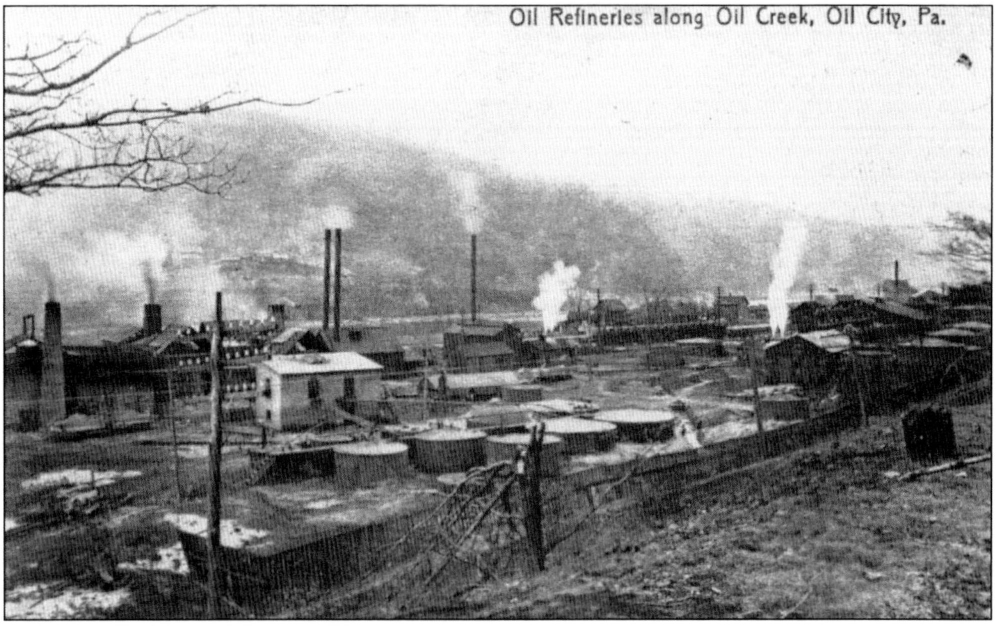

This postcard is captioned "Oil Refineries along Oil Creek"; however, this appears to be an undated image of the refinery at Reno on the Allegheny River. The Empire Refinery was the first to open in Reno in 1879. By 1911, Wolf's Head petroleum products were being made at Reno. Wolf's Head became a part of Pennzoil in 1963.

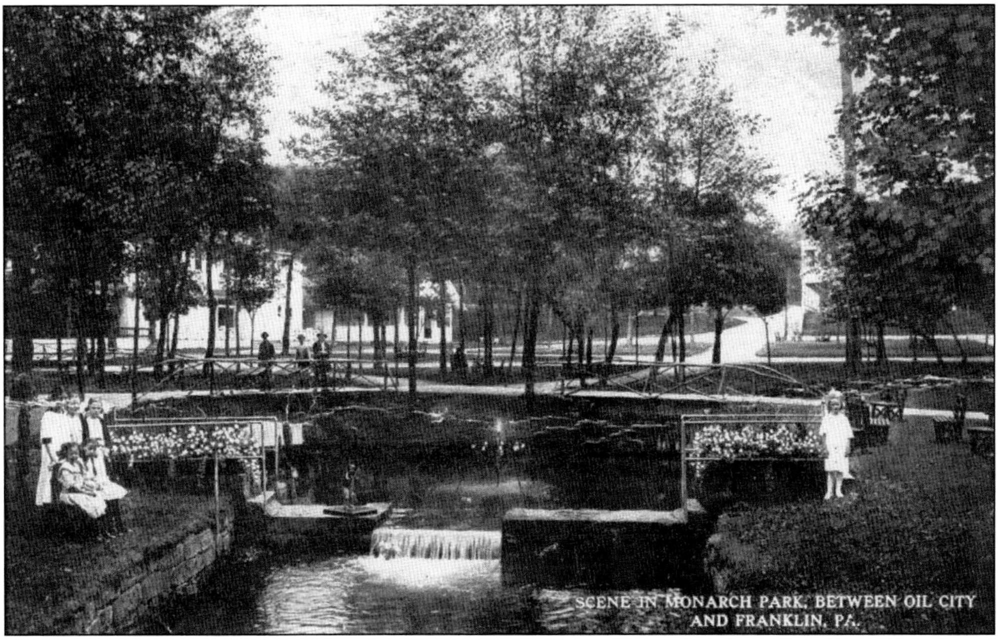

Monarch Park, located between Oil City and Franklin, was a population amusement park in the oil region. The park had a dance hall, roller coaster, bowling alley, and playground for children. Monarch Park was owned by the Citizens Traction Company, which operated a trolley line between Oil City and Franklin. This image of Monarch Park between was postmarked in 1915.

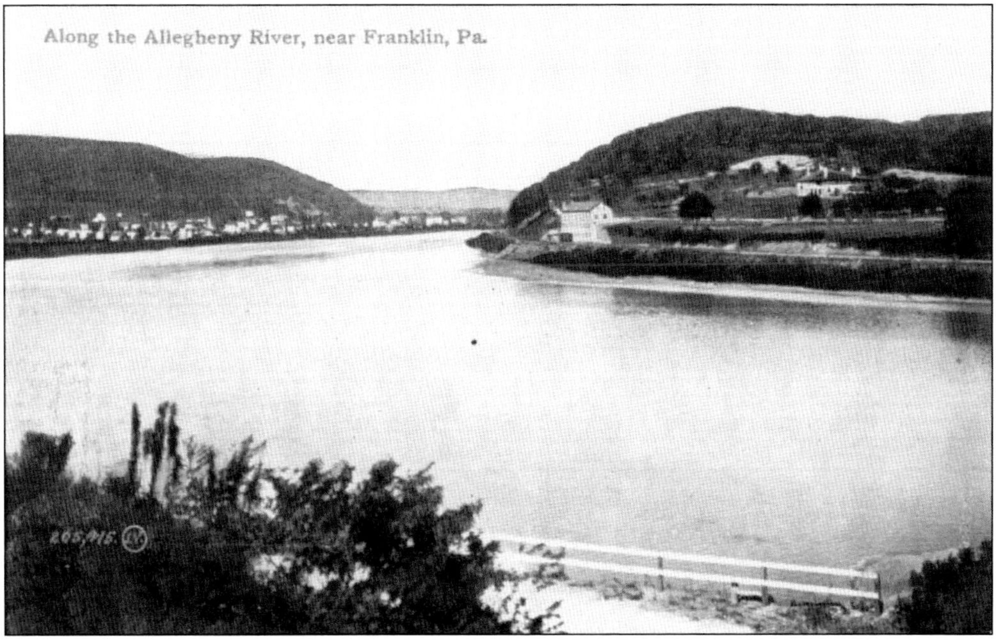

This scene along the Allegheny River just north of Franklin shows a portion of Joseph C. Sibley's River Ridge Farm at the right of the image. Sibley, along with his brother-in-law Charles Miller, founded the Galena Oil Company in Franklin. Sibley was mayor of Franklin in 1879 and later served several terms in Congress as both a Democrat and Republican. He died in 1926.

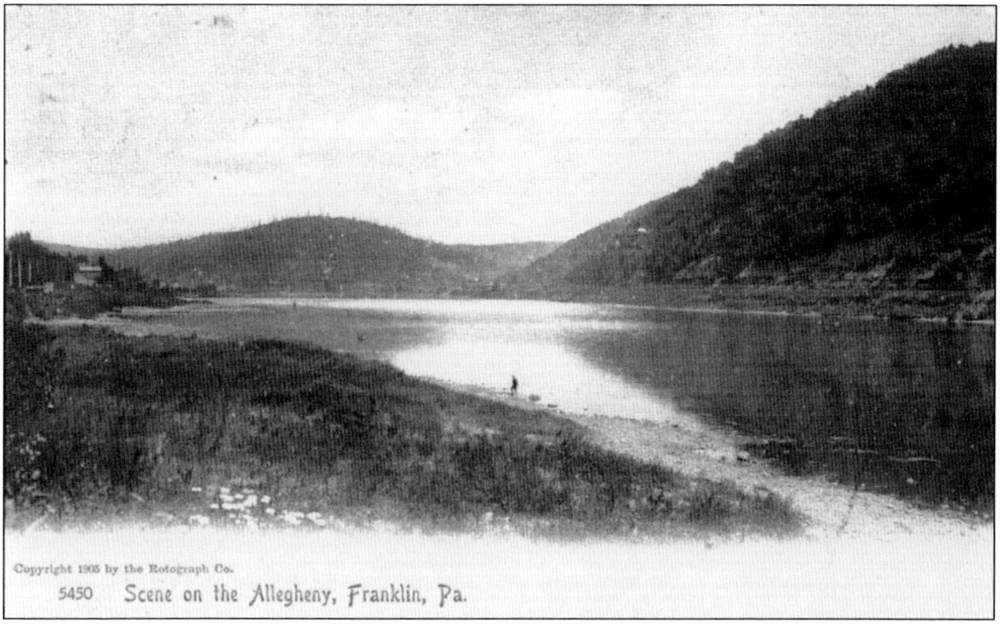

Franklin, surveyed as a town in 1797, is located at the confluence of the Allegheny River with Oil Creek. Franklin was the site of four forts—French, English, and American—that attempted to keep this strategic location secure during the French and Indian War, the American Revolution, and the early days of the American Republic. Among the early commodities traded at the forts was oil, largely sold for medicinal purposes.

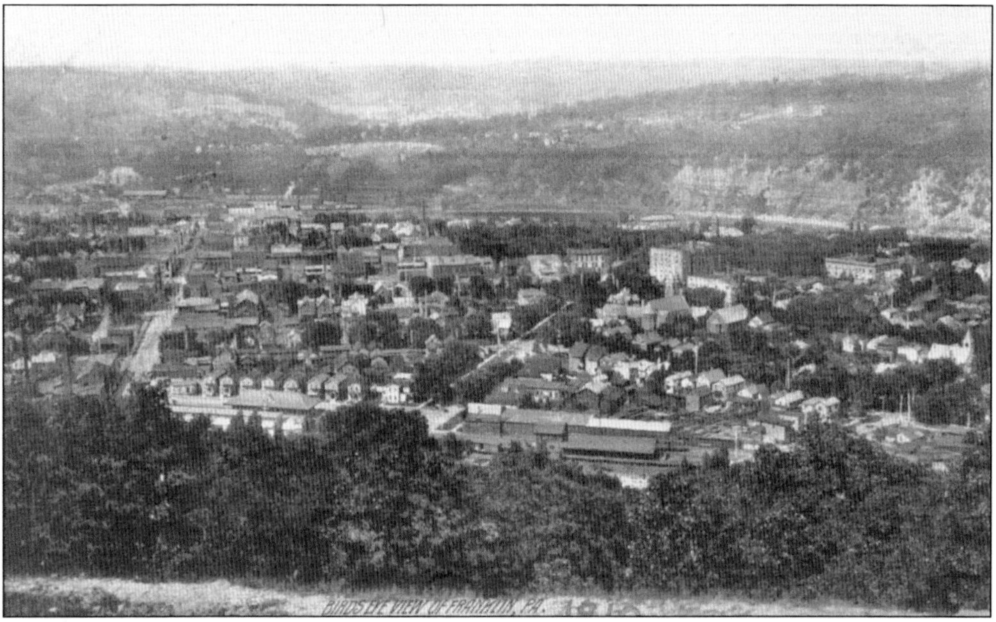

In the early 1900s, Franklin was a residential and manufacturing town with a population of nearly 10,000. Four railroads served Franklin, including the Erie, the Lake Shore and Michigan Southern, the Pennsylvania, and the Franklin and Clearfield. The town had numerous industries, including steel and brass foundries, tool companies, machine shops, and flour mills. This bird's-eye view of Franklin is dated 1918.

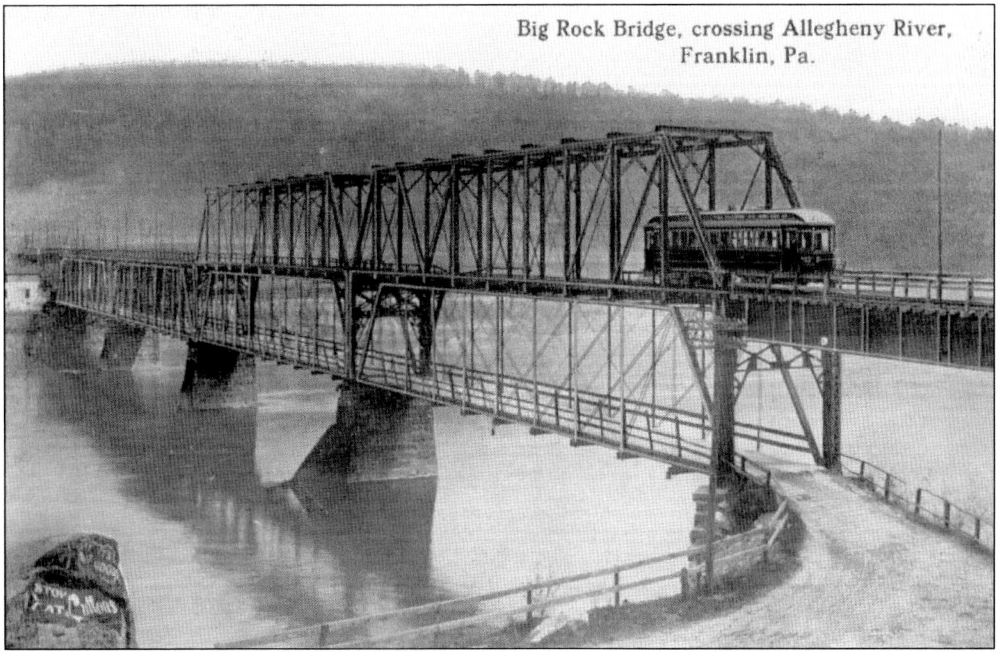

Franklin has a rich history of bridges and crossings over the Allegheny River and French Creek. These postcards show two of these historic crossings. The image above shows the Big Rock Bridge crossing the Allegheny River. A trolley line runs on the upper level. The image below, postmarked 1913, shows the Washington Bridge where the French Creek and Allegheny River meet.

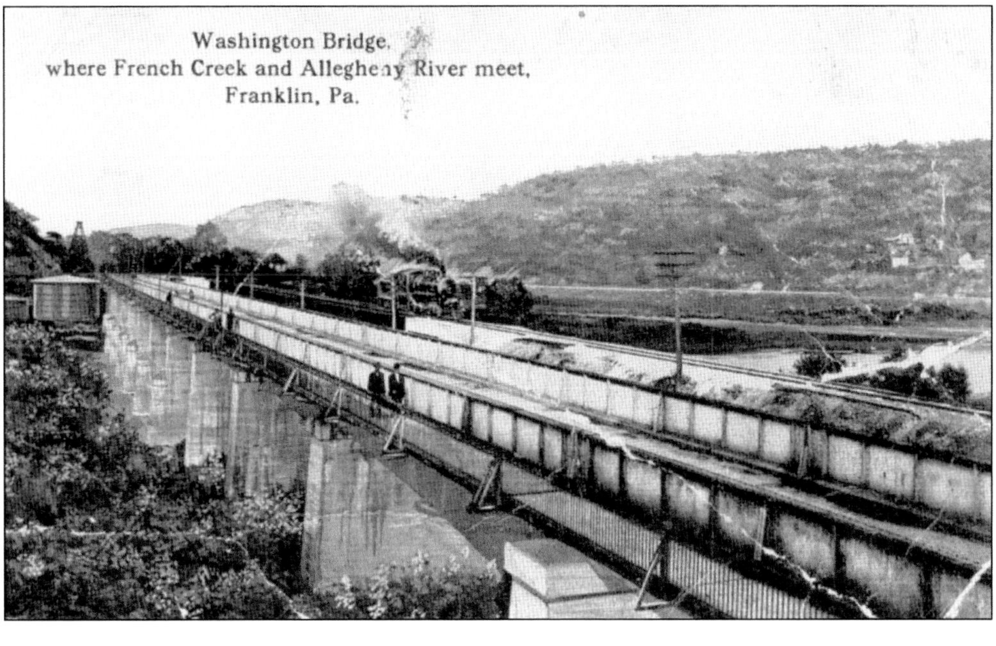

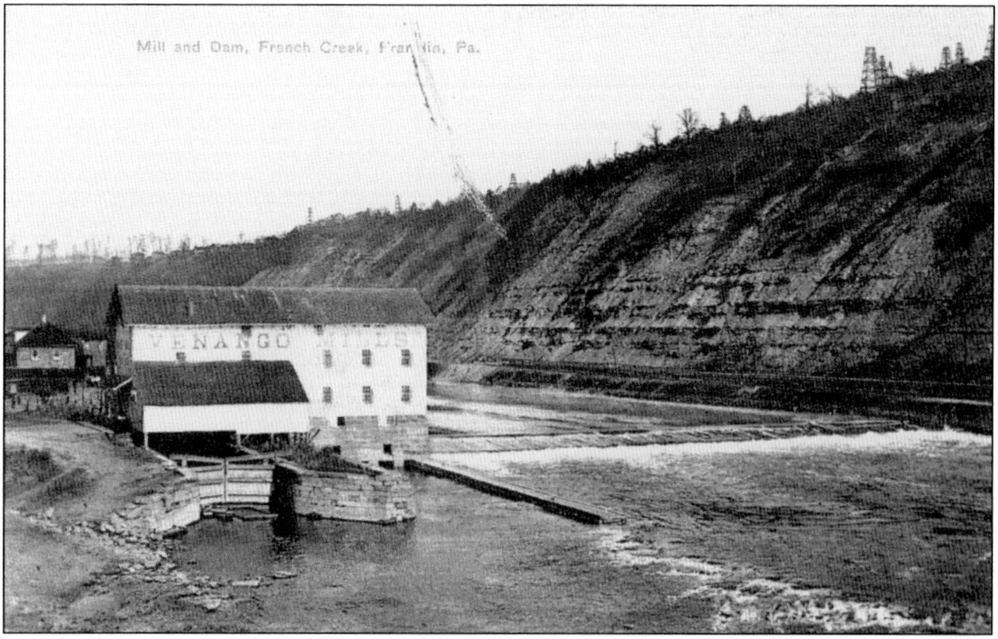

This postcard, postmarked 1910, shows an image of the historic Venango Mills and its milldam on French Creek in Franklin. This image differs little from the lithograph published in the 1865 atlas of the oil region of Pennsylvania, when J. A. Dewar and T. A. Smith were the proprietors. Note the derricks scattered across the top of the valley in this view.

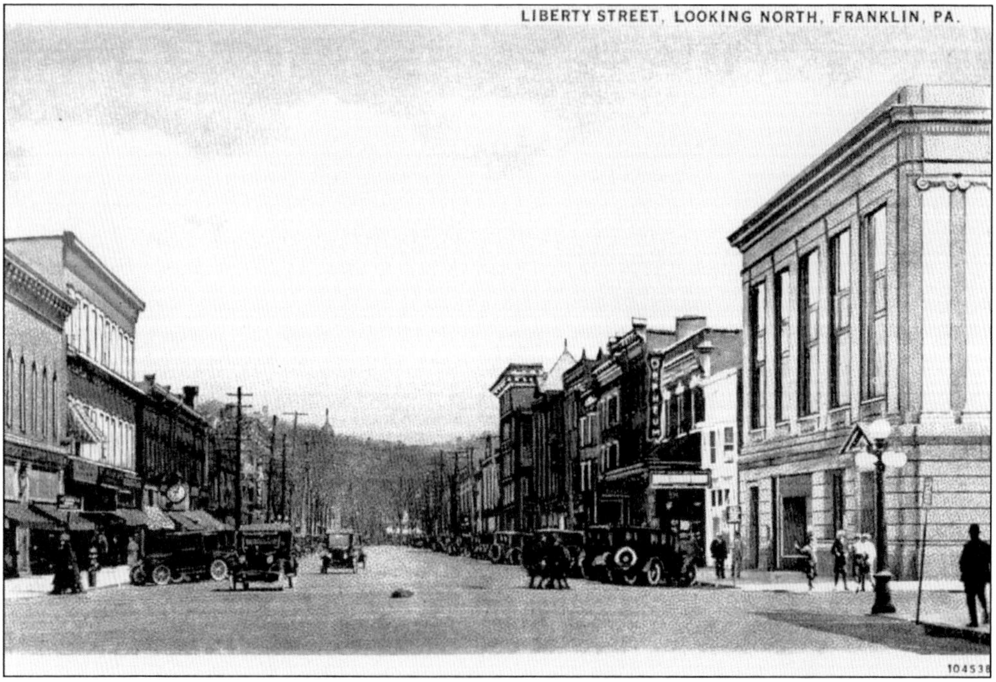

Liberty Street is Franklin's main thoroughfare and commercial center. This view of "Liberty Street Looking North" is from the 1920s. Except for the vehicles, much of the historic downtown of Franklin appears the same today.

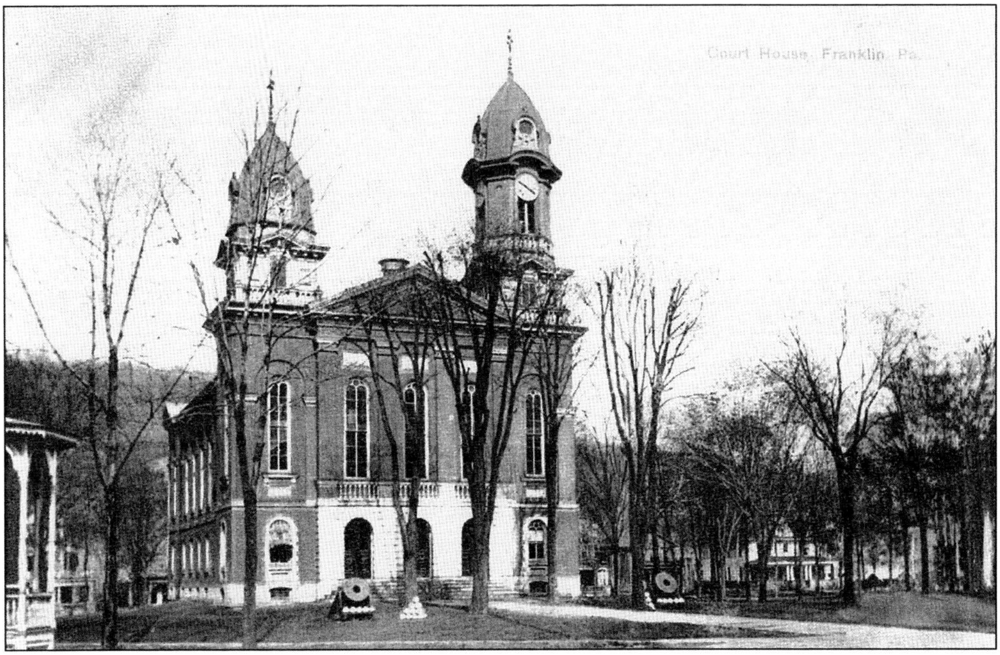

The twin-towered Venango County Courthouse, located on Liberty Street, was designed either by architect Samuel Sloan or Addison Hutton and was built in 1868 and stands today. It was preceded by two other courthouses, the first built in 1811 and the second built in 1848. County commissioners originally wished to fireproof the courthouse of 1848 but found that building a new courthouse was more cost-effective.

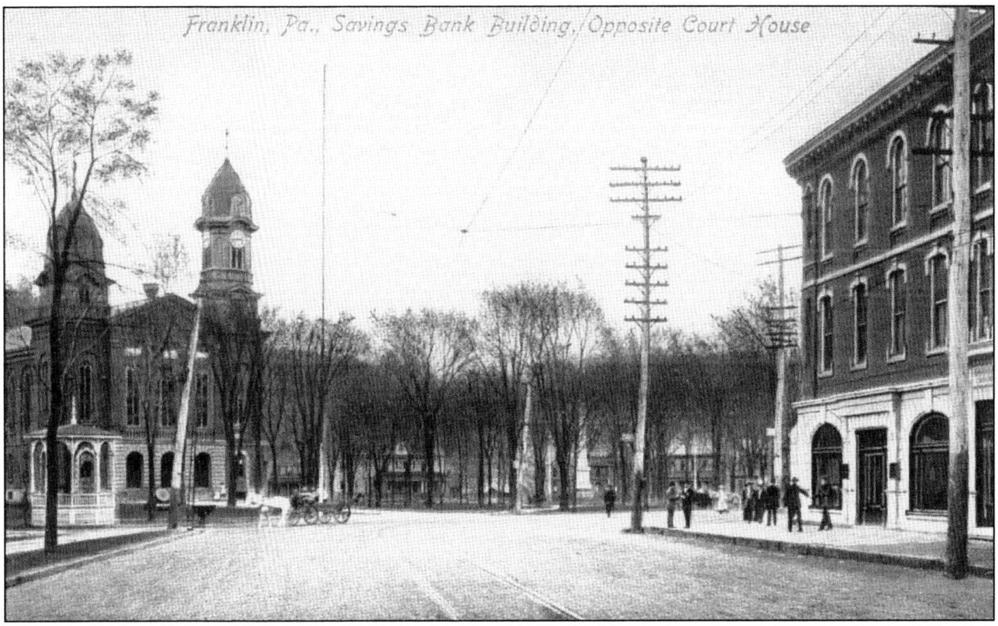

The former Franklin Savings Bank Building was located opposite the courthouse on Liberty Street. This postcard was an advertisement sent to prospective customers in the early 1900s. The message on the card's back reads, "Money. If you wish to save, borrow, lend, or invest, you are invited to come to the Franklin Savings Bank."

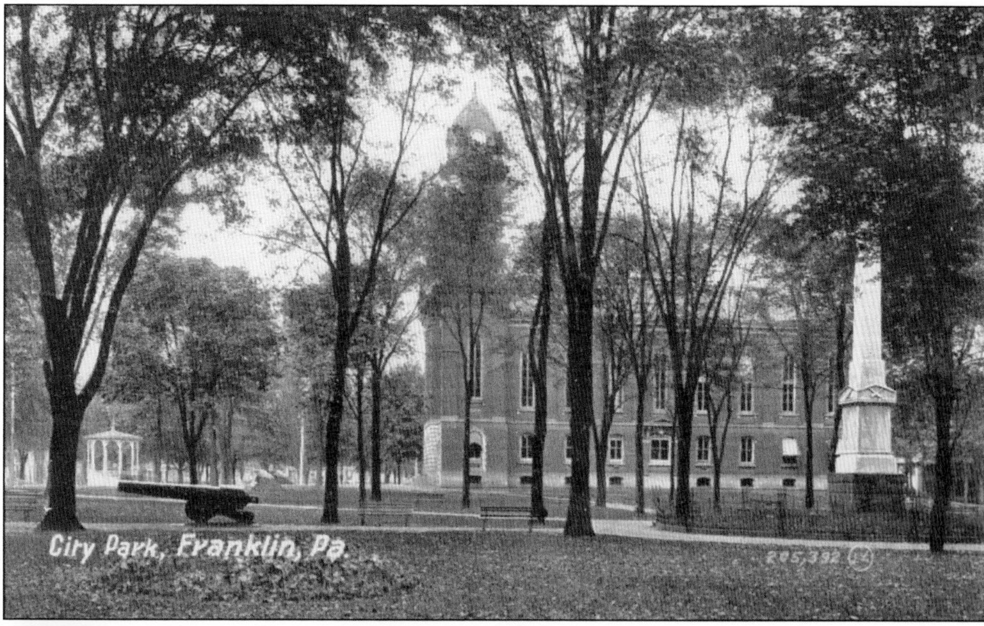

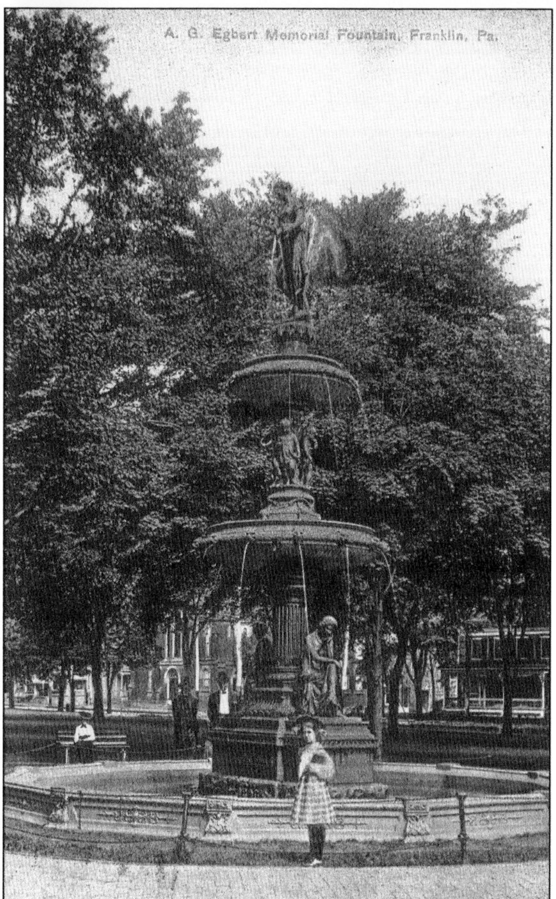

The City Park in Franklin is located adjacent to the Venango County Courthouse. A focal point of the park is the A. G. Egbert Memorial Fountain, shown on the postcard below, which was postmarked in 1910. A. G. Egbert was a strong supporter and promoter of the park, and the fountain was dedicated as a memorial for his work.

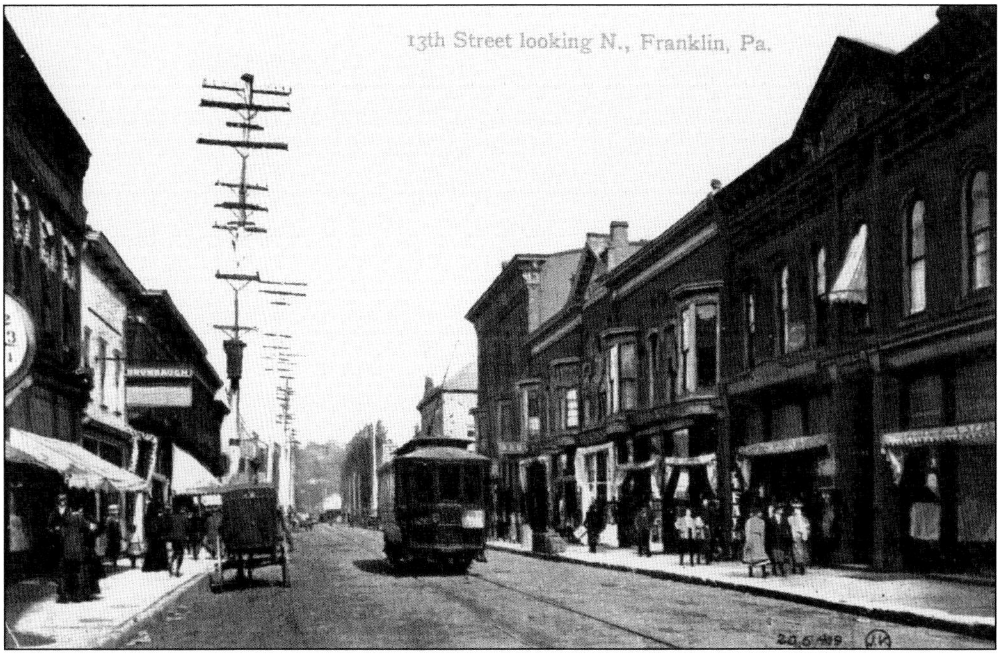

These two postcards show different views of Franklin's Thirteenth Street looking north. The images are undated but are from the early 1900s. Note the trolley running down the street in the image above. Trolleys provided local service in Franklin and also connected the town with Oil City eight miles away. The image below shows some of the former storefronts on Thirteenth Street, including Brumbaugh's Furniture and Carpets on the left side of the street.

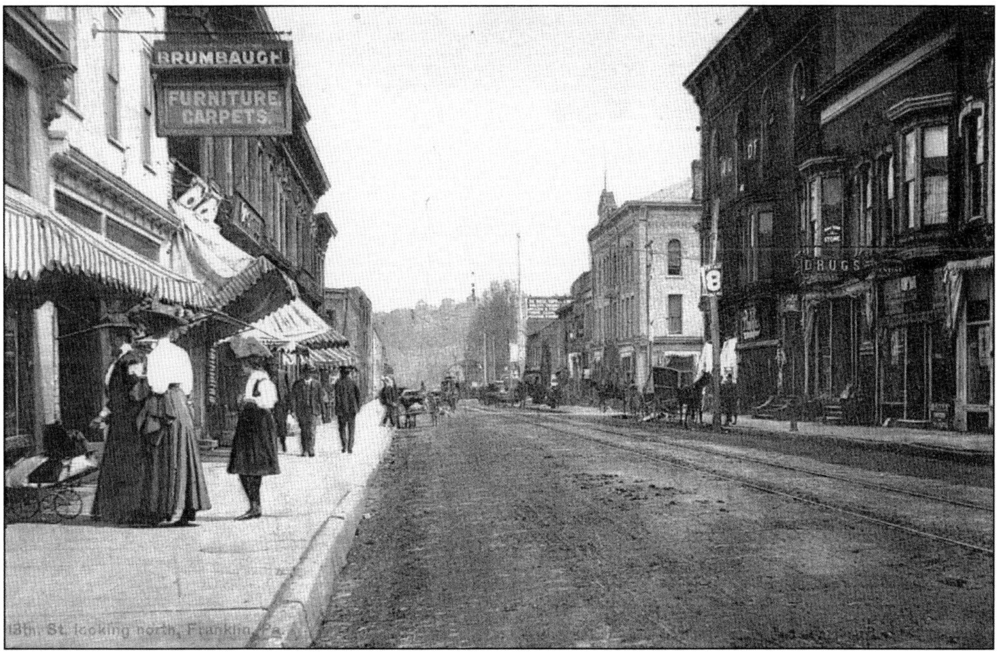

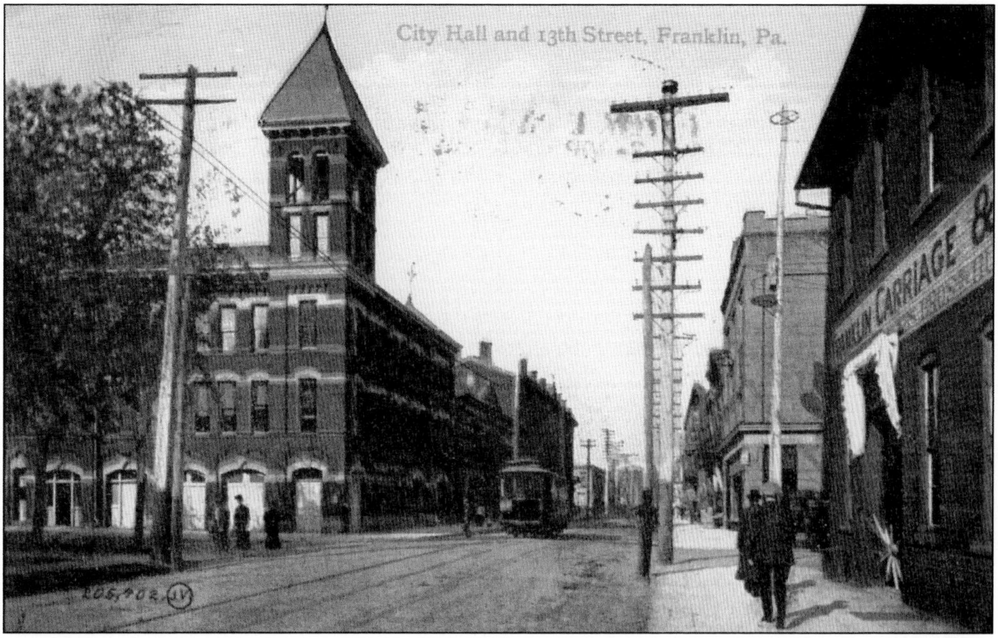

The former city hall in Franklin, as seen in this postcard postmarked 1913, was located on Thirteenth Street. The ubiquitous trolley, the local transport of the day before the rise of the automobile, passes by the building in this image.

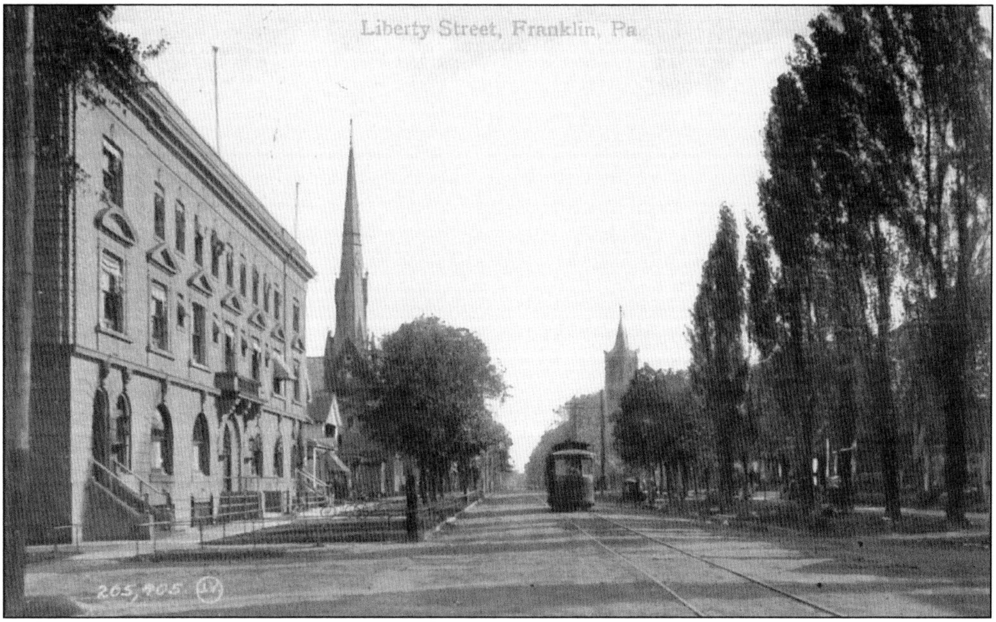

This view of Liberty Street in Franklin shows a trolley about to pass the Galena-Signal Oil Company Building. The company's name came from the lead sulfide, called galena, used in the refining process, which gave the oil unusual endurance as a lubricator. Charles Miller and John Coon were the driving forces in the growth of Galena Oil. In 1878, Galena Oil was sold to Standard Oil with Miller as the Galena Division head.

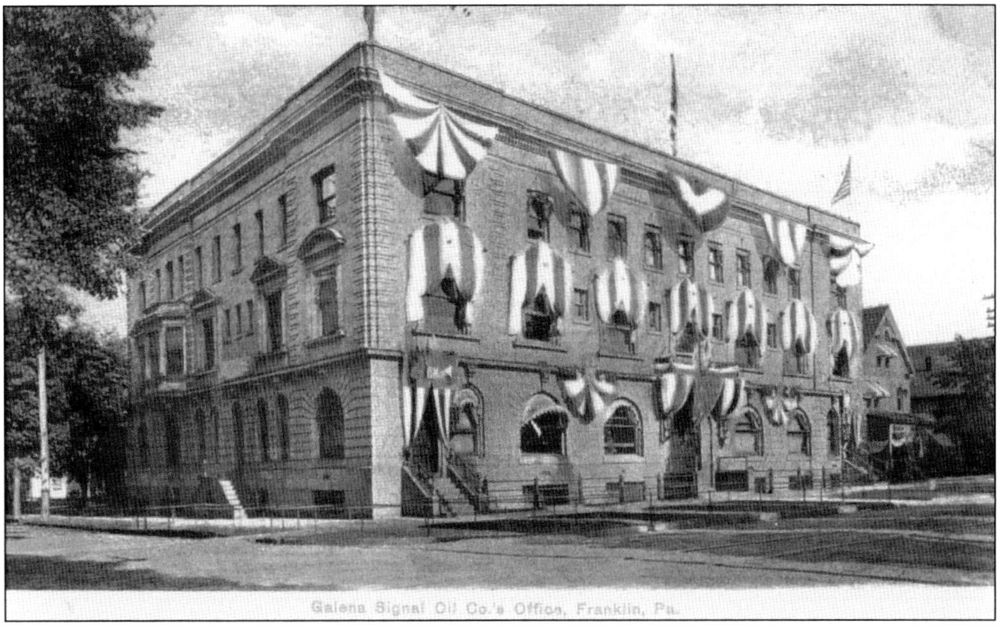

The Signal Oil Works was formed in 1879 after sale of Galena to Standard Oil. Joseph C. Sibley was chairman, and Charles Miller was principal stockholder. In 1901, Galena Oil and Signal Oil Works formed the Galena-Signal Oil Company, and in 1911, the company became independent on the breakup of Standard Oil Company. Valvoline Oil absorbed the company in 1931, and operations in Franklin ceased by the 1940s.

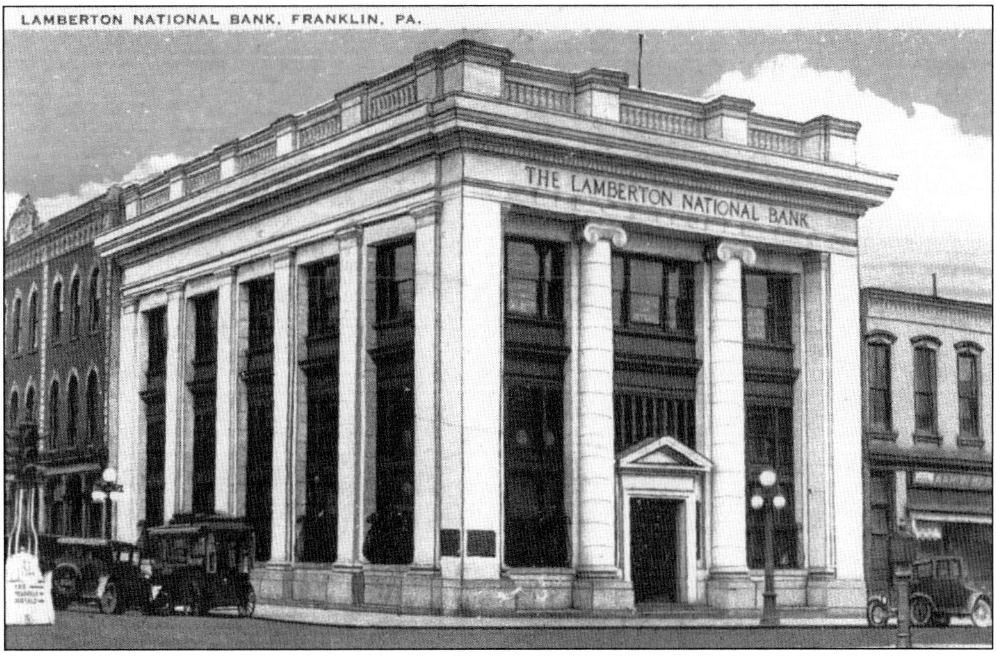

Robert Lamberton opened one of the first banks in the oil region in Franklin in 1859. The Lamberton National Bank at the corner of Liberty and Thirteenth Streets was one of the hallmark buildings in Franklin in the early 1900s. This postcard shows a view of the Lamberton National Bank in the 1920s.

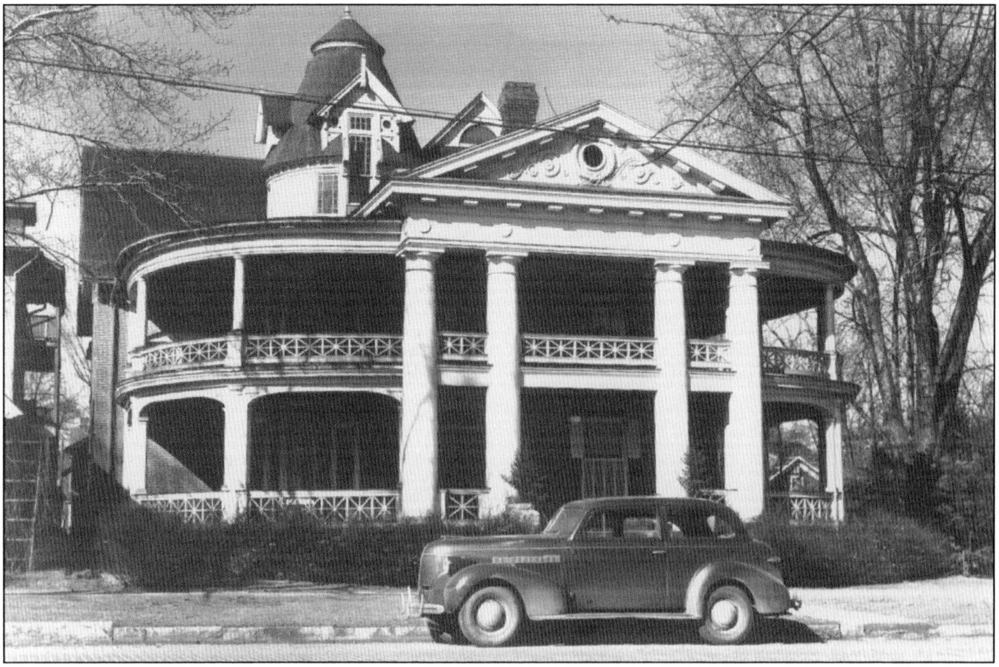

One of the icons of Liberty Street in Franklin is the Franklin Club, seen here in a late 1940s–era image. The Franklin Club was once an important place for socializing in the oil region frequented by doctors, lawyers, and other well-connected people.

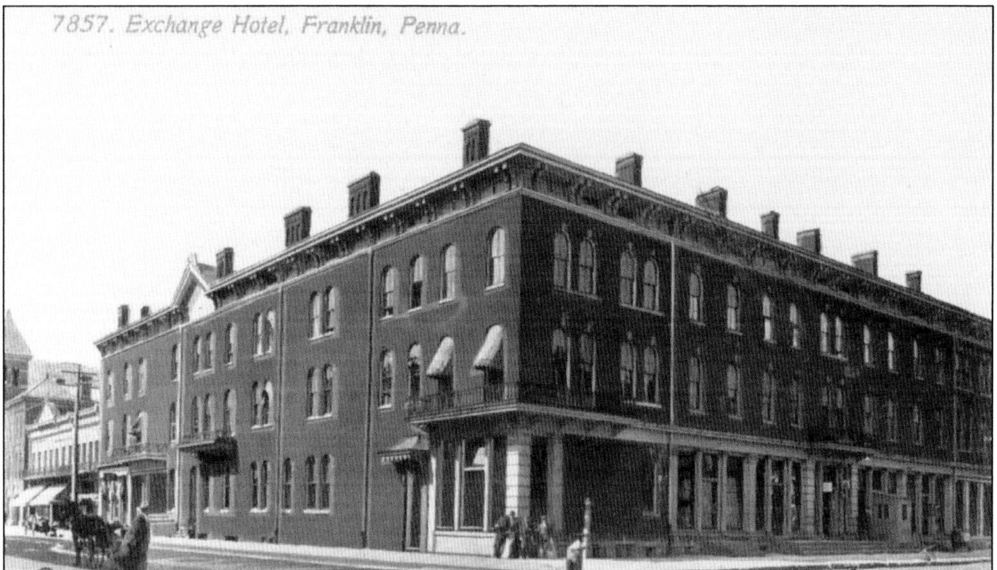

The Exchange Hotel in Franklin once stood at the intersection of Liberty and Thirteenth Streets. In 1871, Pres. Ulysses S. Grant stayed at the Exchange Hotel during his tour of the oil region. In a speech to hotel guests, Grant noted that the oil region was vital to the North winning Civil War: "Pennsylvania contributed her quota during the rebellion but also her oil products as an element of national wealth and resource."

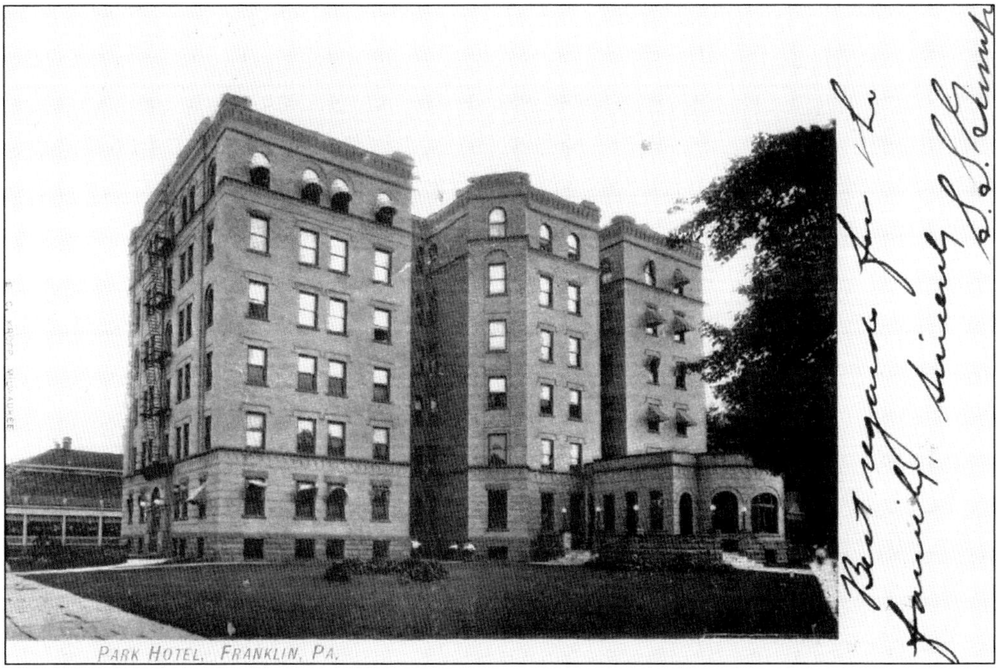

The Park Hotel was one of several hotels in and around Franklin catering to businessmen and visitors to the oil region in the early 1900s. This postcard of the Park Hotel was postmarked in 1907.

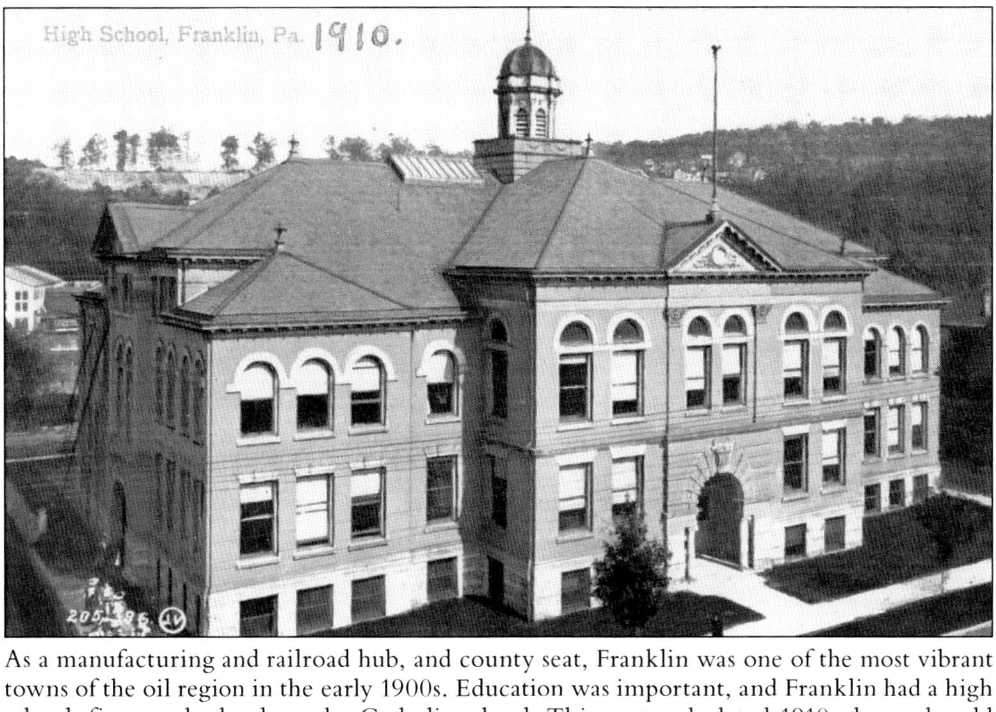

As a manufacturing and railroad hub, and county seat, Franklin was one of the most vibrant towns of the oil region in the early 1900s. Education was important, and Franklin had a high school, five ward schools, and a Catholic school. This postcard, dated 1910, shows the old high school building in Franklin.

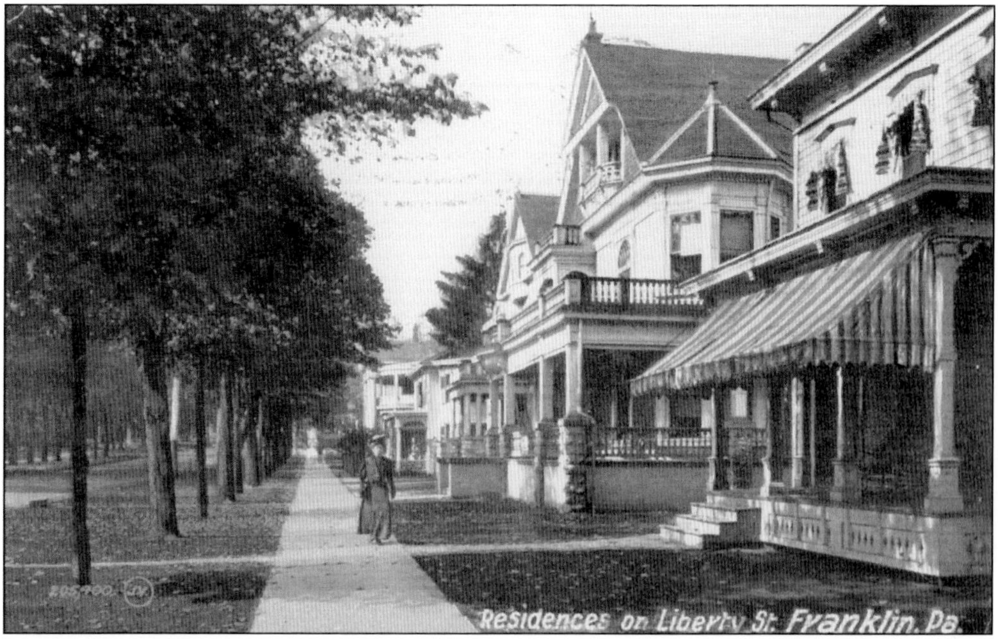

Franklin is well known for its diversity of architecture, especially of its residences. The postcard above, postmarked in 1910, shows residences on Liberty Street, where many Victorian houses stand today. The postcard below shows a view of Miller Park, a fine residential area in Franklin built on the former "Galena Hill." Many of the town's prominent citizens lived in Miller Park.

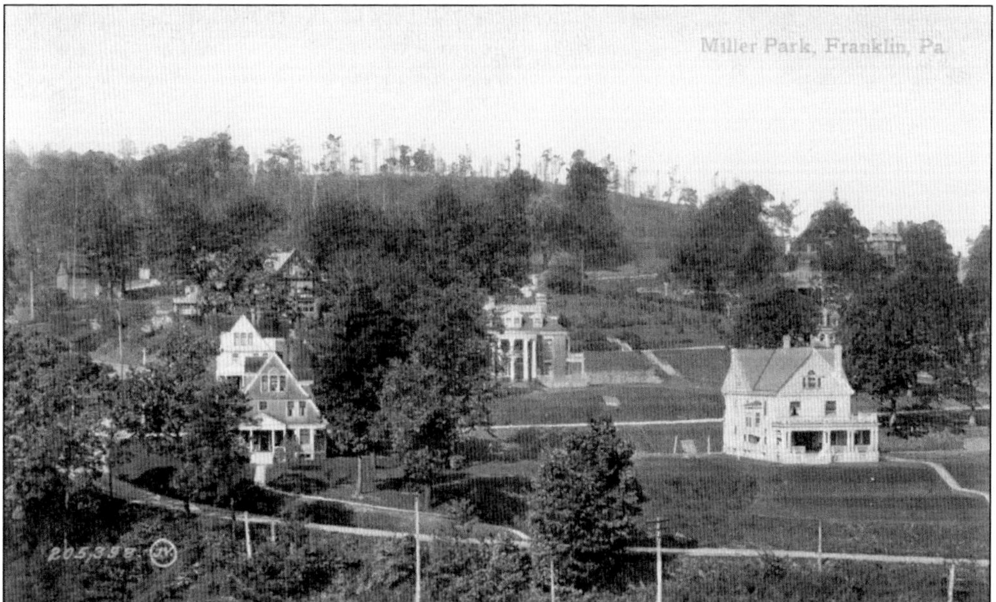

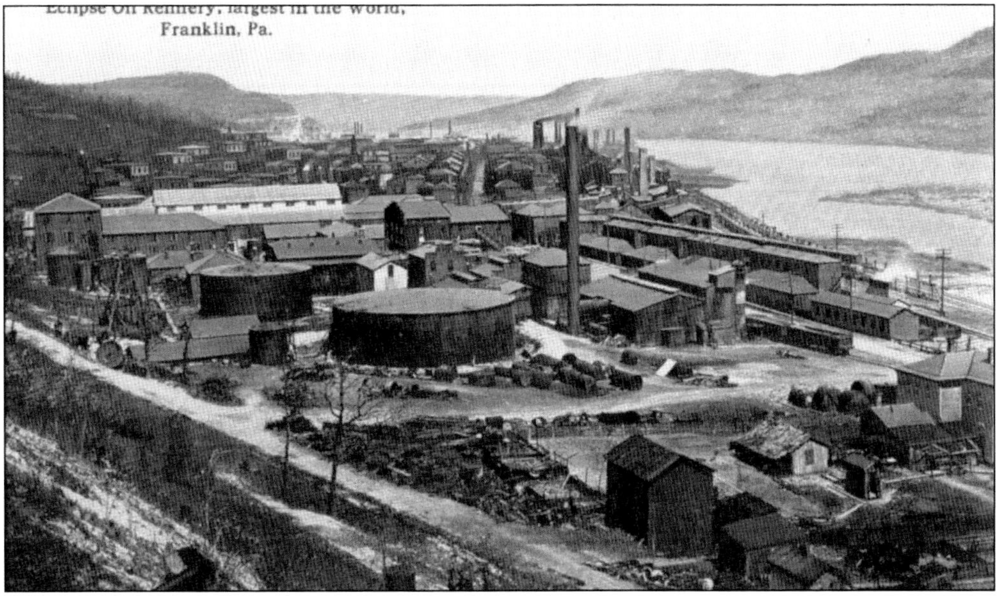

The Eclipse Refining Company, established on the Allegheny River north of Franklin in 1872, was "the largest of its kind in the world" by 1877. Eclipse Oil Refinery was the financial hope of local producers who wished to stand up to the Standard Oil Company's monopoly. The refinery, however, fell into bankruptcy within four years. Standard Oil Company bought Eclipse Oil's refinery and turned it into a productive enterprise. In 1904, the refinery covered 125 acres, processed 10,000 barrels of crude daily, and produced 130 different commercial petroleum products.

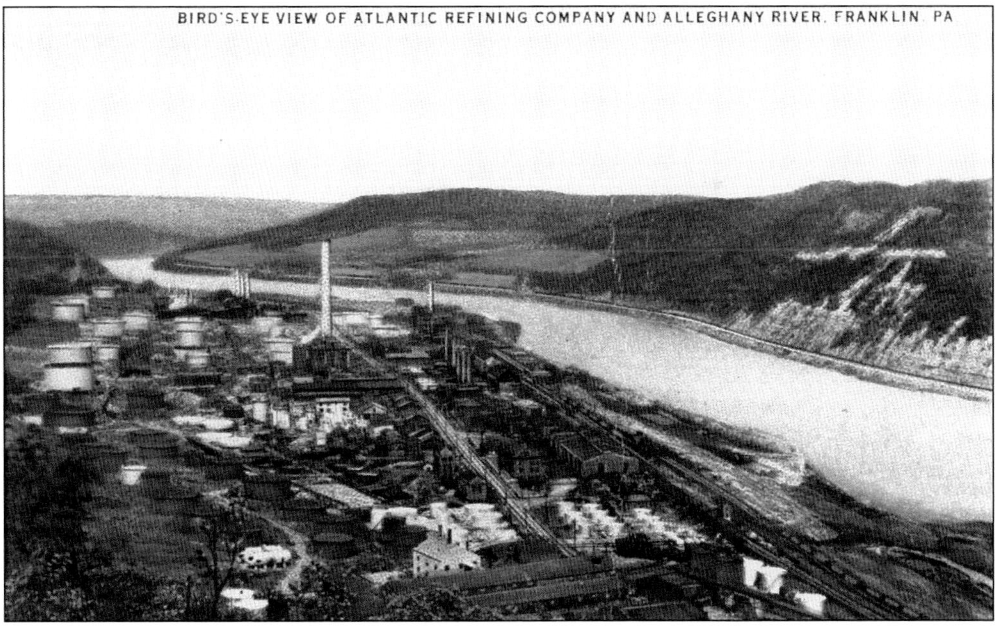

In the muddy history of the Eclipse Refining Company, Eclipse became a division of the Atlantic Refining Company within the Standard Oil Company. When the Standard Oil Trust was broken up in 1911, Eclipse remained within Atlantic Refining. This image is a bird's-eye view of the Atlantic Refining Company at Franklin, the former Eclipse plant.

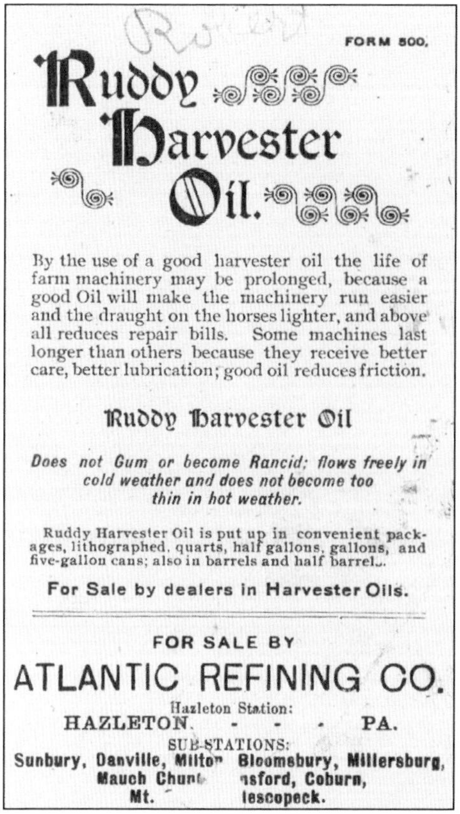

Ruddy Harvester Oil was a product of the Atlantic Refining Company, the company that assumed control of Eclipse Refining after the Standard Oil Company breakup. Eclipse was well known for producing quality lubricating oils for farm machinery as well as many other commercial petroleum products. Atlantic Refining Company continued in the footsteps of Eclipse in producing Ruddy Harvester Oil and other lubricating products.

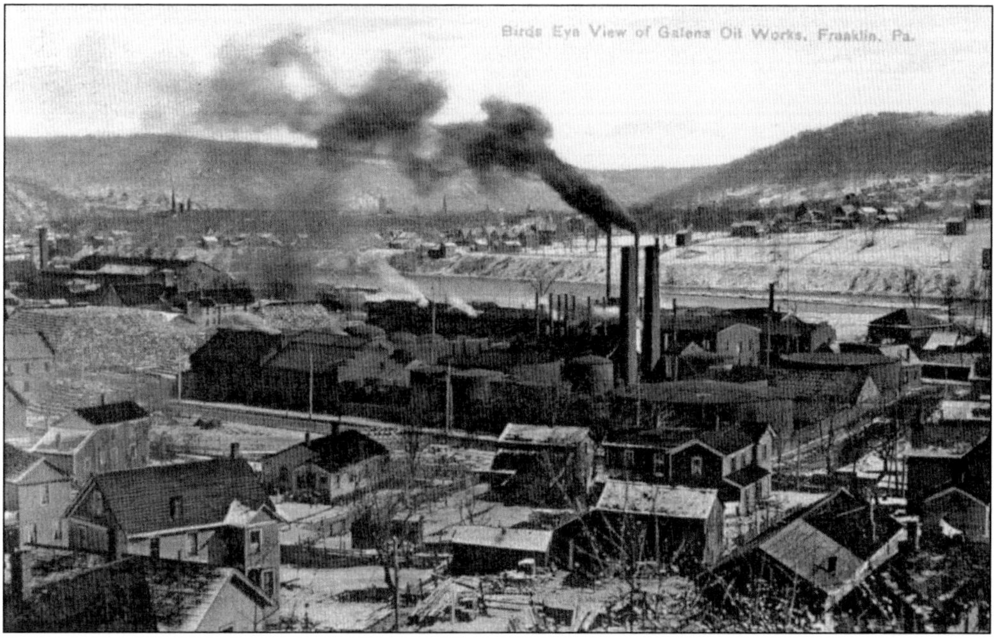

The Galena Oil Refinery was relocated to the old Dale Refinery on French Creek in Franklin after the original factory burned in 1870. Galena was sold to Standard Oil in 1878. By 1895, Galena oils lubricated 95 percent of the railroad mileage in North America and Mexico.

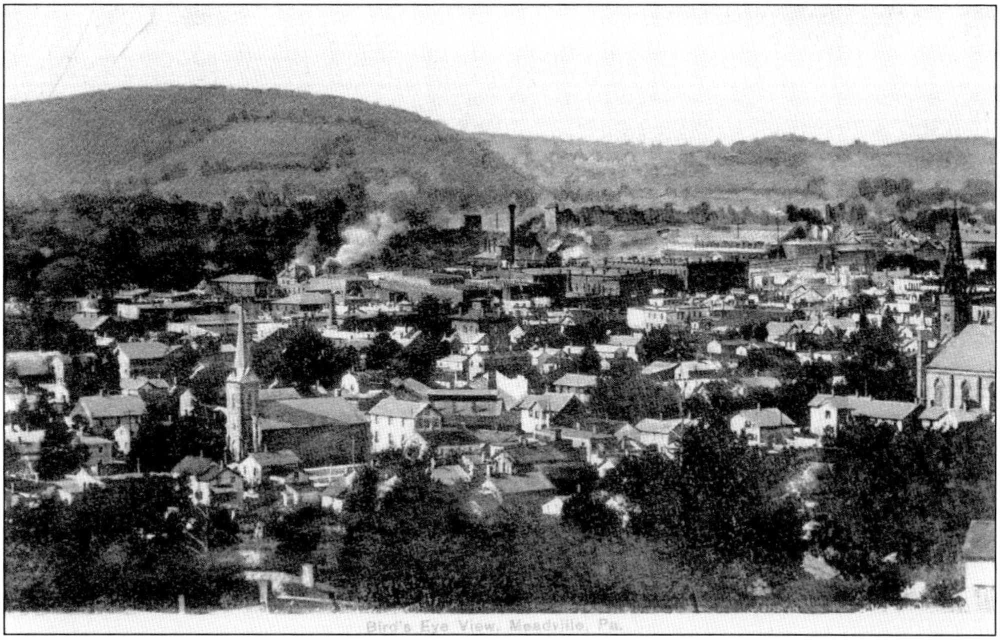

Meadville in the early 1900s was largely a residential community with few industries. The Phoenix Iron Mills and the Meadville Malleable Iron Company were the largest industries. Boiler, vice, tool, and corset works were located in the town along with a distillery, a brewery, a sawmill, and a gristmill. The landscape around Meadville was largely agricultural.

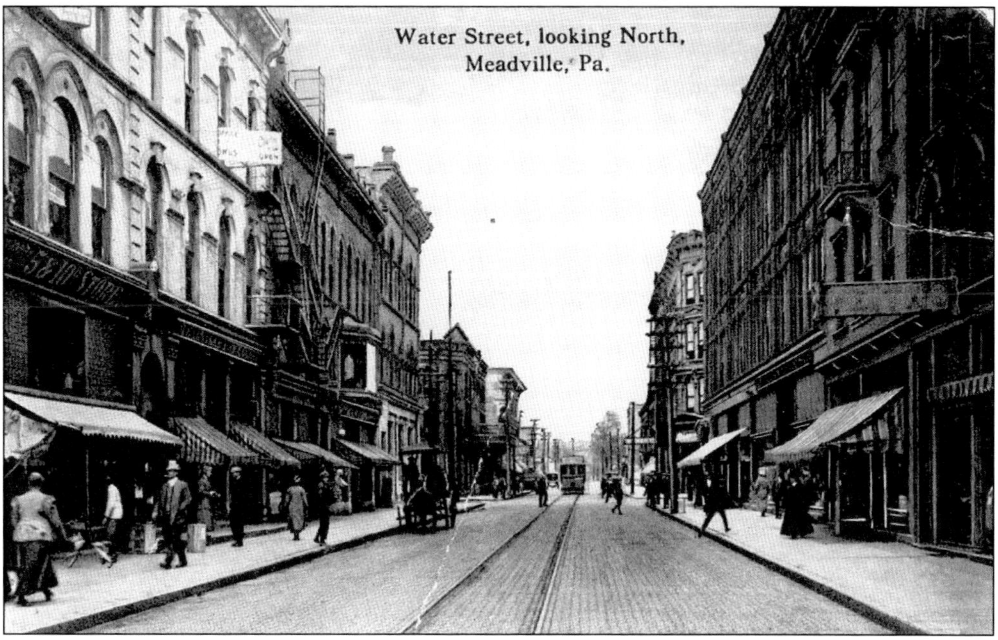

Meadville was another one of the gateway towns into the oil region. An 1870 treatise on the oil region noted that "A large proportion of the trade [at Meadville] comes from the Oil Region . . . It is well supplied with all the business branches requisite, banks and banking houses, three newspapers . . . churches and schools. Its hotels are numerous." This view of "Water Street looking North" in Meadville is from the early 1900s.

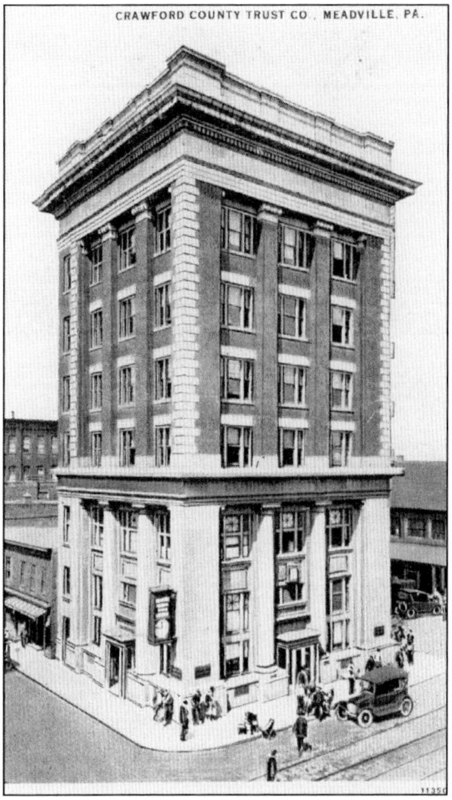

Among the prominent buildings in Meadville is the Crawford County Trust Company. The *Pennsylvania Farmer*, published in Meadville, lists this advertisement for the Crawford County Trust Company in 1908: "An old and well established Savings Department with over Four Thousand Active Accounts. The Peoples Bank. Four per cent paid on all Savings Accounts. Call and get one of our Home Banks. They are loaned Free, or write for particulars."

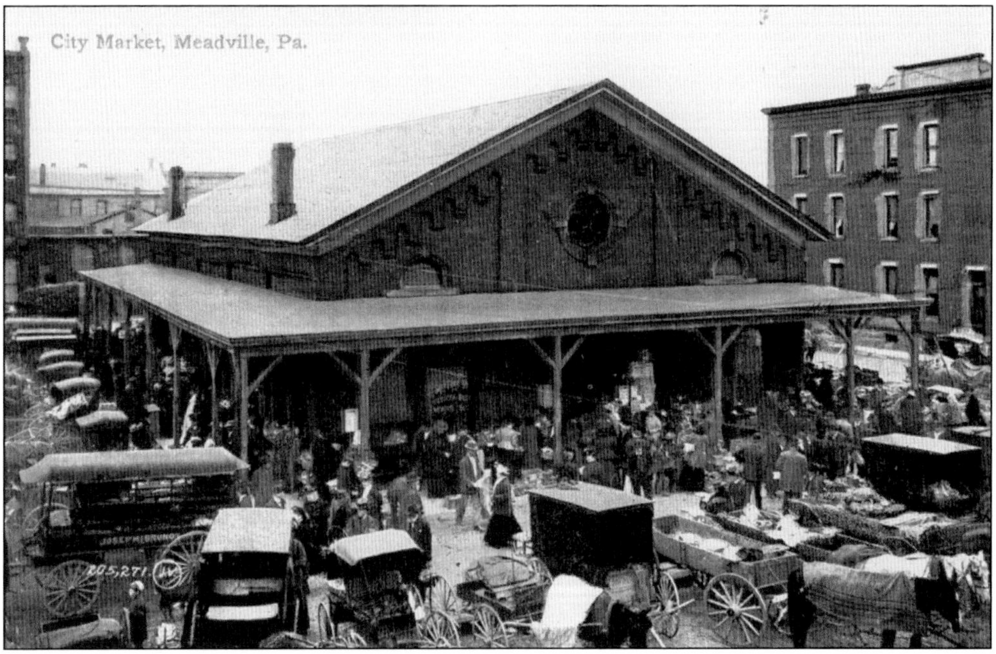

The City Market was built in Meadville in 1870 by the town council. Today the market building houses local vendors and the Meadville Council on the Arts. The message on the back of this postcard, postmarked 1911 reads, "It is nice here but oh so dirty and smoky."

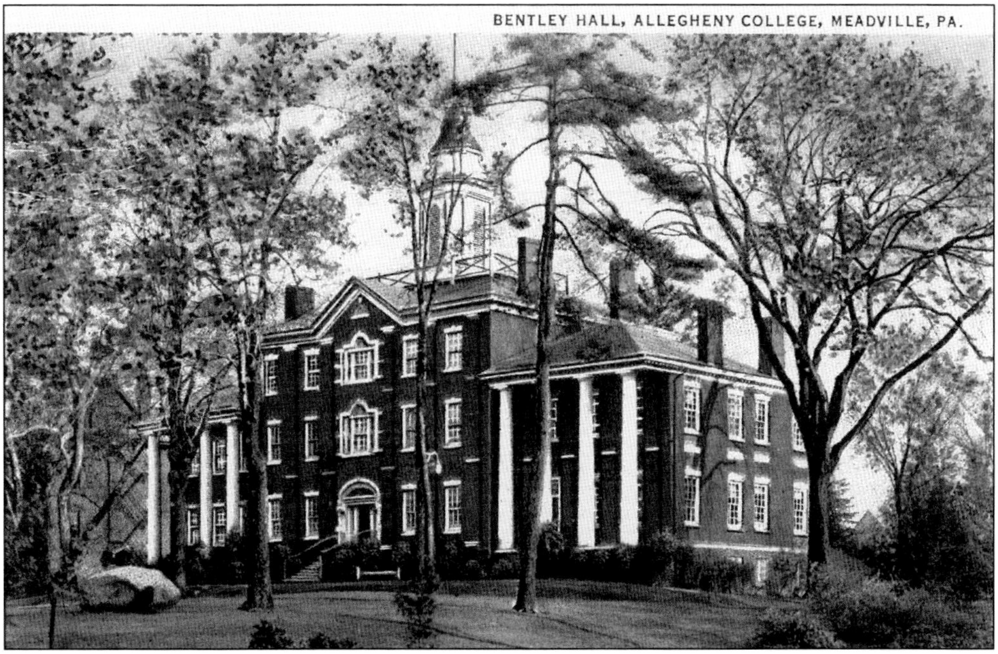

Allegheny College was home to one of the oil region's foremost historians, Dr. Paul H. Giddens. Dr. Giddens was on the faculty at Allegheny College from 1931 to 1953, becoming chair of the history and political science department. He was also curator of the Drake Well Museum from 1943 to 1953. Dr. Giddens published 86 articles and 5 books on the oil industry, including the *Early Days of Oil* and *Pennsylvania Petroleum 1750–1872: A Documentary History*.

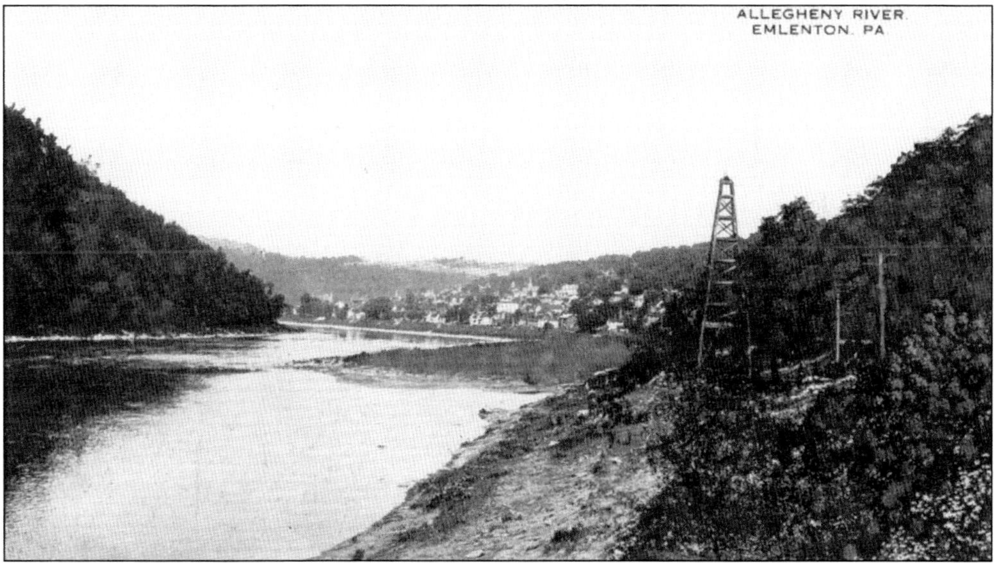

Downstream from Franklin and French Creek is Emlenton in Venango County. Oil speculators moved south along the Allegheny River in the late 1860s, drilling wells from Franklin to Foxburg in Clarion County. *New York Daily Tribune* correspondent J. R. G. Hazard traveled down the Allegheny River from Franklin and noted the activity: "from Franklin to Emlenton there is a succession of these small wells [pumping four to five barrels daily], the aggregate yield of which is about 1,000 barrels a day."

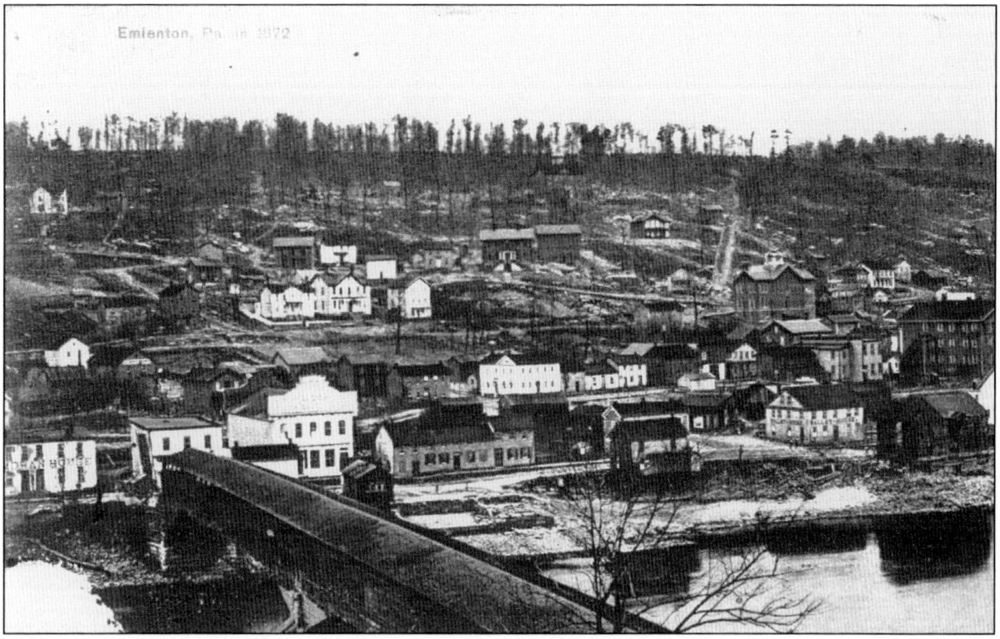

These two postcards show the changes in Emlenton from 1872, above, to 1940, below. The above image shows Emlenton just when the oil boom had hit. The wooden covered bridge across the Allegheny River seen in this image was built in 1856. It was lost in the flood of 1883. The image below shows a modernizing Emlenton about to enter World War II. The message on this card reads, "Nice little town, apple blossoms, lilacs, tulips."

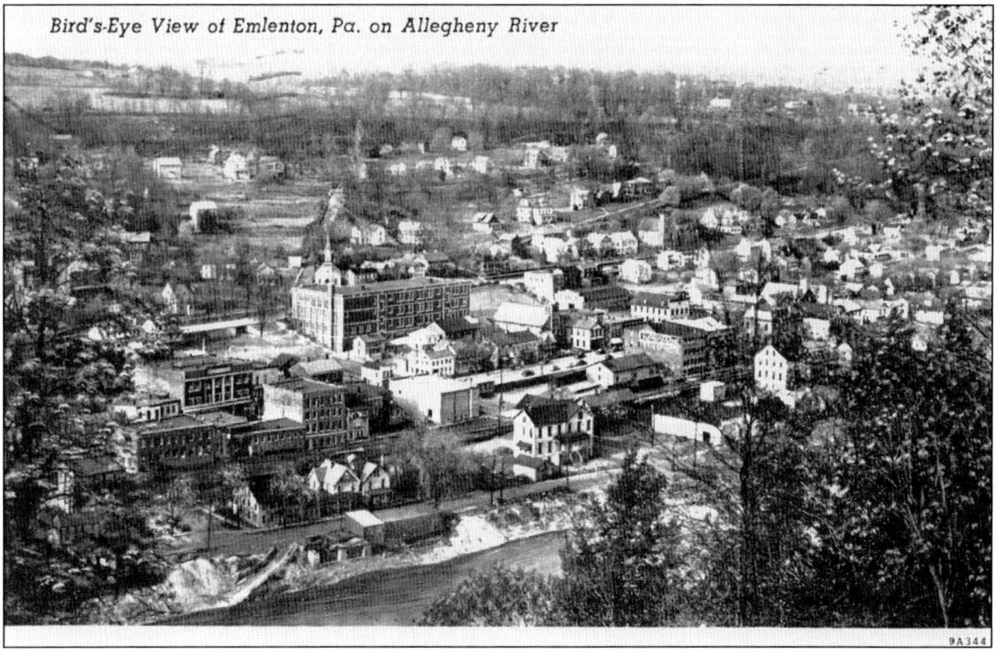

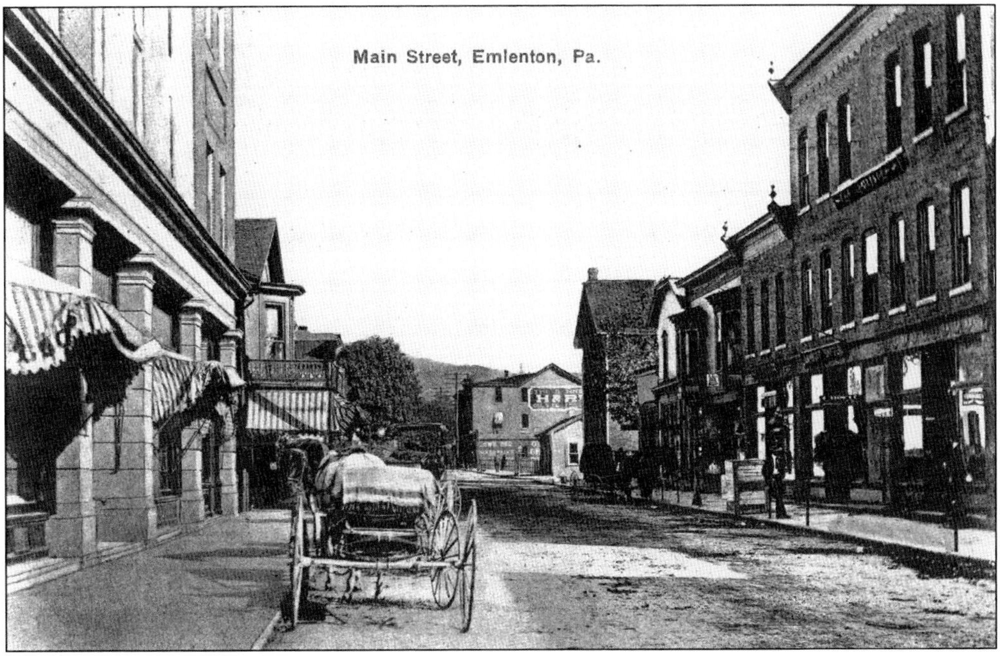

Emlenton had a population of about 1,300 in the early 1900s. The principal industries of the time were the Emlenton Refinery, a woolen mill, and several small machine shops. These two postcards of the time show views of Main Street in Emlenton, the transportation dominated by horse-drawn wagon and carriages. The upper card was postmarked in 1910.

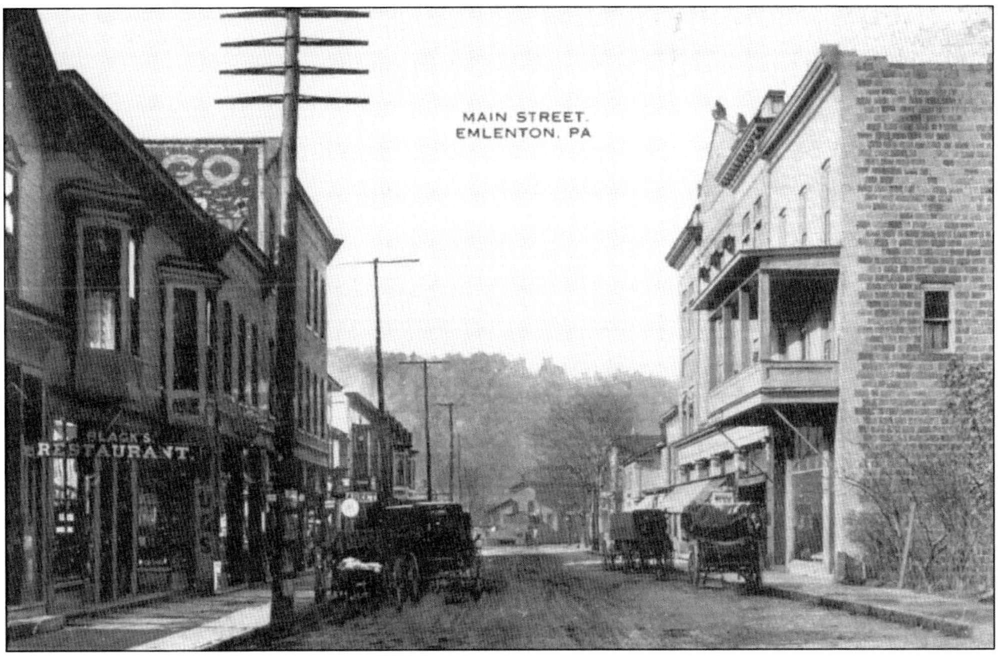

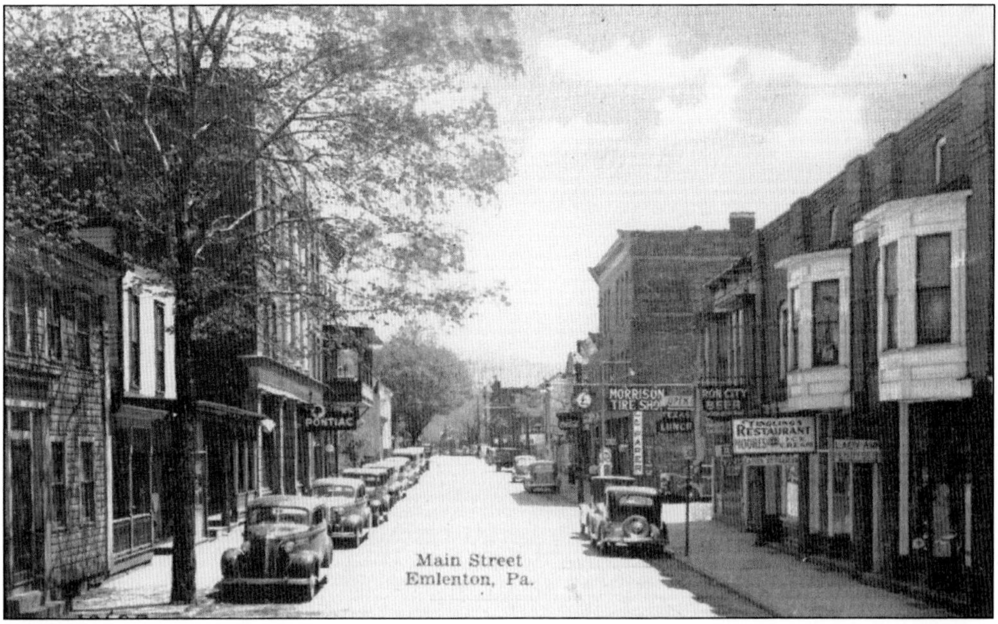

This Silvercraft postcard shows a view of Main Street in Emlenton from the 1940s. Among the businesses and products seen in this image are the Pontiac dealer at the image's left, the Yingling Restaurant serving Moore Ice Cream to the right, and the Morrison Tire Shop and Texas Lunch also on the right. IC (Iron City) beer appears to be the drink of choice at one establishment.

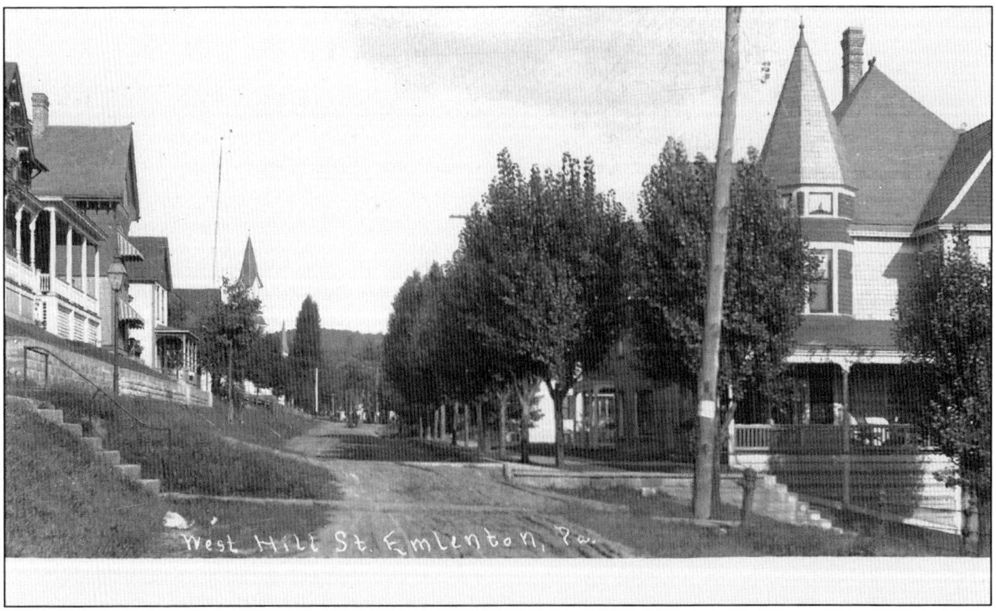

Like Tidioute and Franklin upstream on the Allegheny River, Emlenton developed a fine residential Victorian architecture that persists to today. This image of West Hill Street, postmarked in 1907, shows a residential area in Emlenton with a mix of housing styles—Victorian and otherwise—and churches in the distance.

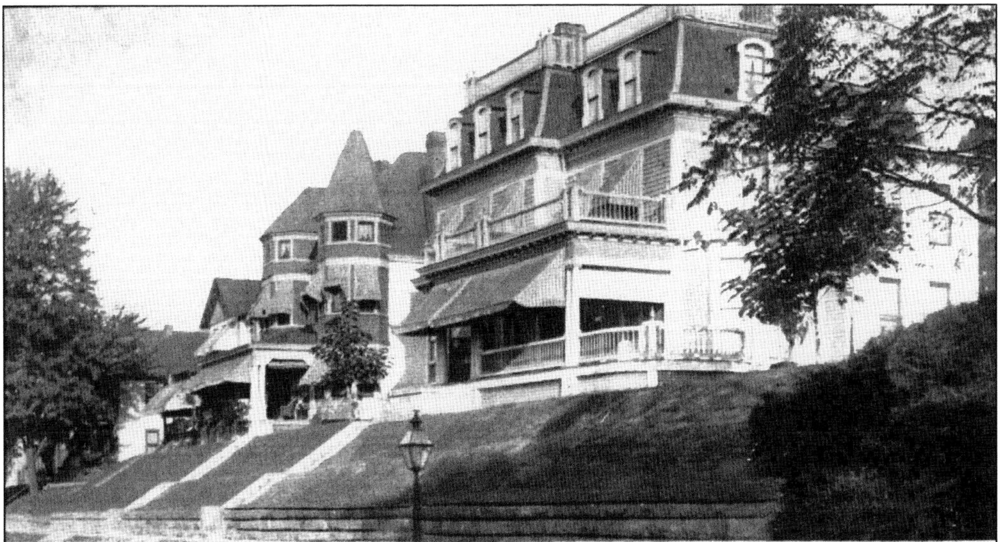

Crawford Residence, Hill Avenue, Emlenton, Pa.

The Crawford family of Emlenton had a strong interest in the oil industry. Among the most influential of the family was Dr. A. W. Crawford, who served two terms in the state legislature and was appointed as U.S. consul to Belgium by Abraham Lincoln in 1861. Dr. Crawford worked hard to market Pennsylvania oil in Europe for lighting purposes, once proclaiming, "Let the oil springs and wells of the Allegheny Valley profitably let their lights shine in this direction."

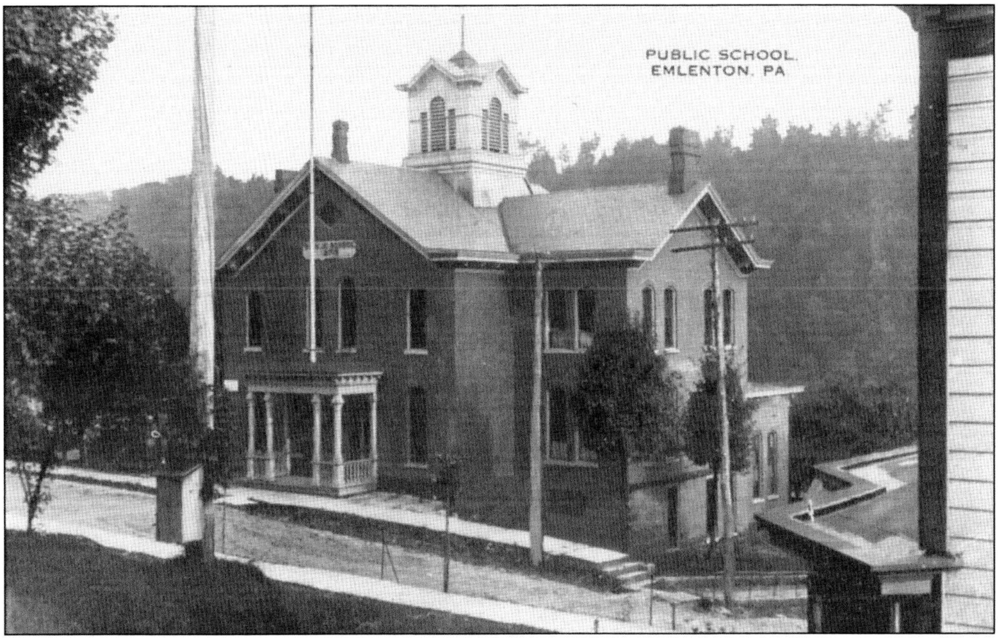

Oil historian J. J. McLaurin had great things to say about Emlenton and its citizens in his 1898 book *Sketches in Crude Oil*. McLaurin wrote, "Emlenton is supplied with natural gas and is noted for good schools, good hotels and get-up-and-get citizens and is wide-awake in every respect." This image of a public school at Emlenton shows one of the good schools McLaurin praised in his book.

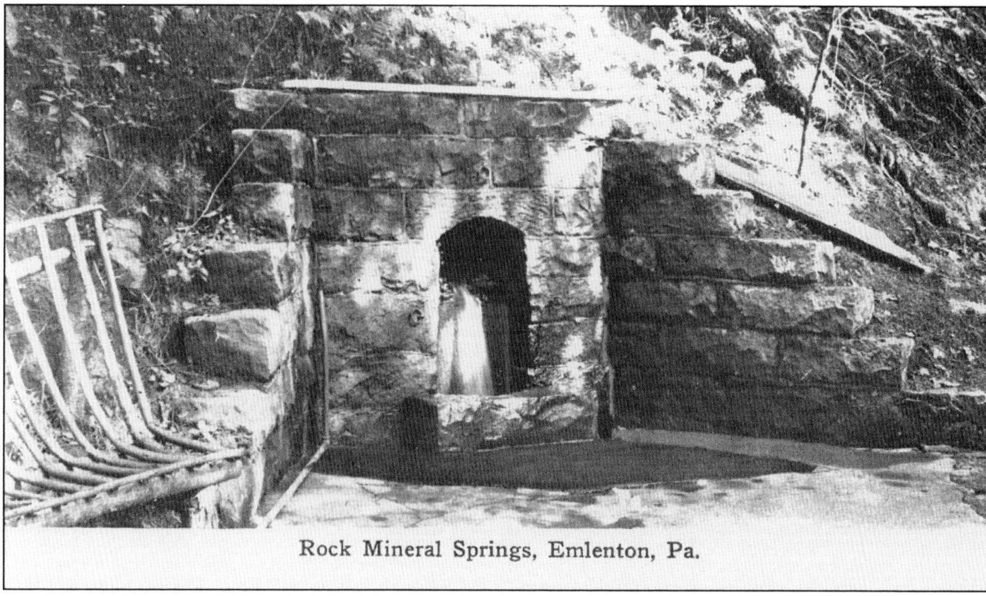

Ebenezer Crawford drilled the Ritchey Well along Ritchey Run at Emlenton in 1867. The well, still in production, is now part of Mineral Spring Park, which is owned by the Borough of Emlenton. The Ritchey Well is maintained in working condition as a living history of oil production. This postcard, postmarked 1908, shows Rock Mineral Springs for which the park is named.

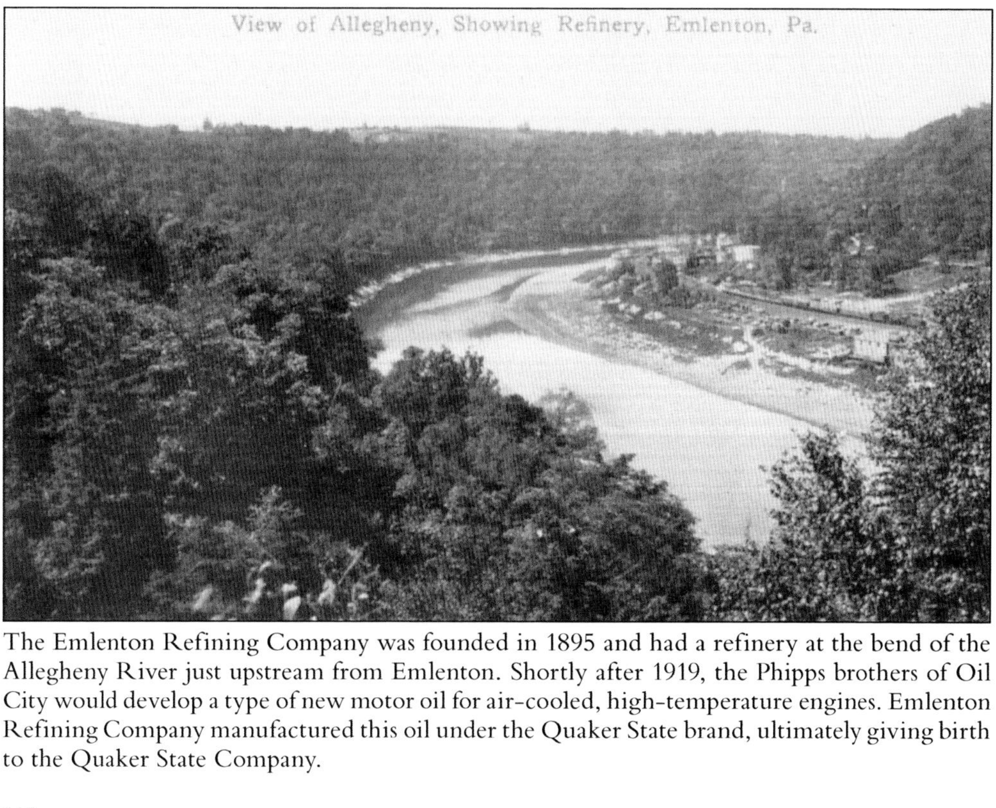

The Emlenton Refining Company was founded in 1895 and had a refinery at the bend of the Allegheny River just upstream from Emlenton. Shortly after 1919, the Phipps brothers of Oil City would develop a type of new motor oil for air-cooled, high-temperature engines. Emlenton Refining Company manufactured this oil under the Quaker State brand, ultimately giving birth to the Quaker State Company.

Eight

Blazes, Floods, Torpedoes, and Gushers

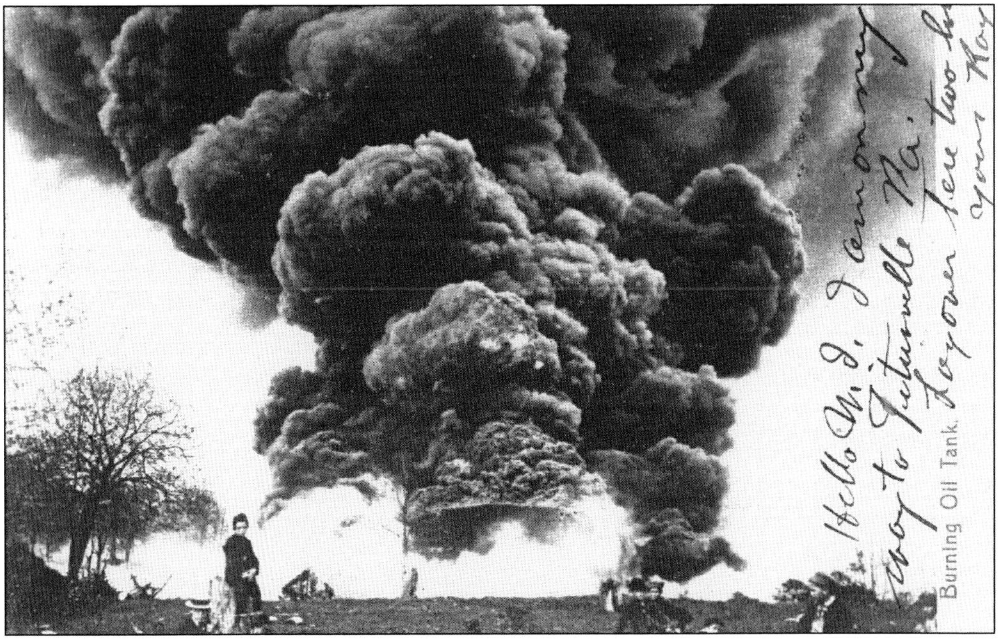

Fires were not an uncommon event in the oil region, from the fire at Drake Well in October 1859 to Titusville's disastrous "Black Friday" of June 11, 1880, when almost 300,000 barrels of oil were ablaze. Lightning was often the ignition source for storage tank fires in the oil region, like the blaze shown here. Postcards showing burning storage tanks with clouds of smoke were quite popular, such as this card with spectators watching the blaze, postmarked 1907.

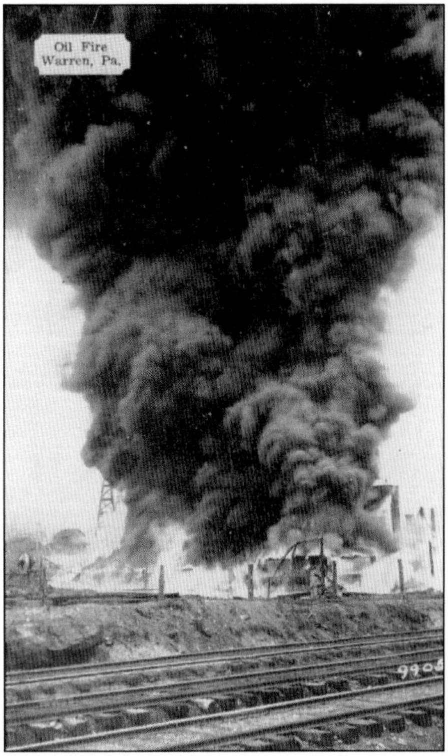

Lightning was often the ignition source for storage tank fires in the oil region, like the blaze shown here. Railroad sparks and carelessness also started their share of fires. This postcard, postmarked 1939, shows a flaming oil storage tank at Warren.

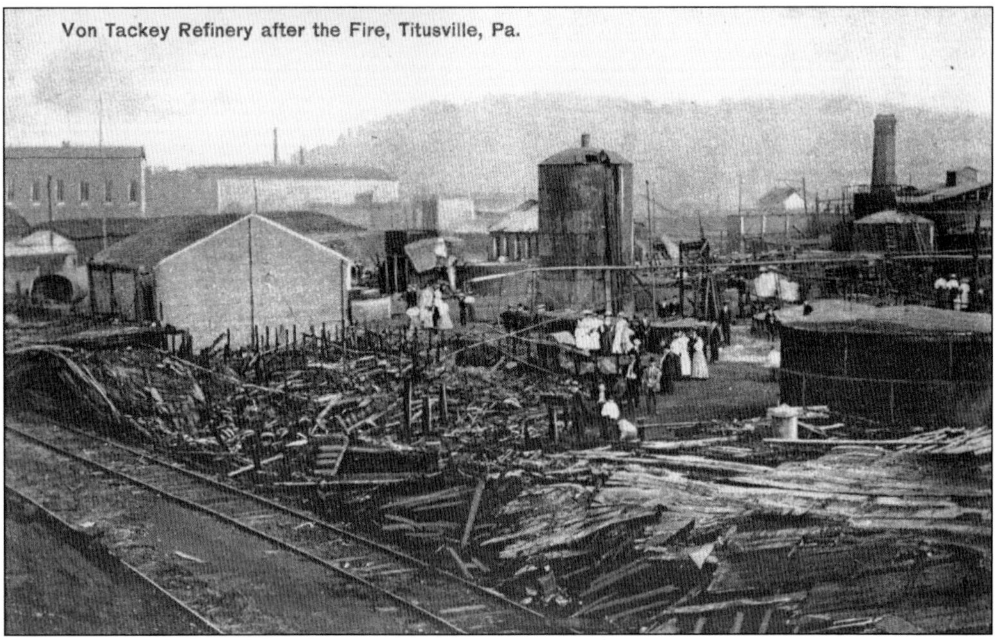

The Titusville Oil Works or "Von Tackey [*sic*] Refinery," shown on this postcard postmarked 1906, was one of the many refineries in and around Titusville in the late 1800s and early 1900s. It was in business until 1945, headed by Frank von Tacky. The image of the refinery after "the fire" likely refers to the flood and fire that hit Titusville on June 4, 1892, when a dam failure upstream caused a torrent of water to hit the town.

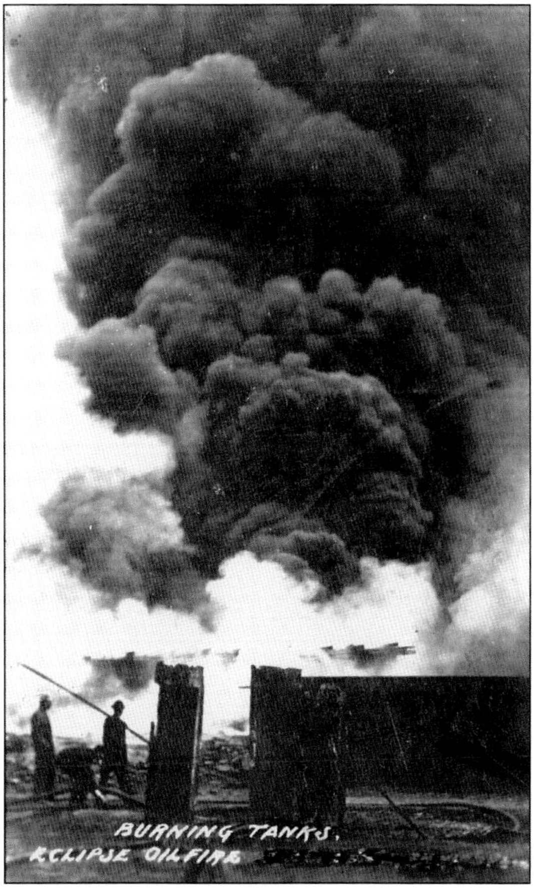

As refineries of the oil region grew in size and output, setbacks like fires of small to moderate extent were handled quickly to minimize losses to productivity. The Eclipse Oil Refinery on the Allegheny River near Franklin suffered a blaze on October 11, 1911, that destroyed some buildings and burned a few oil storage tanks. The refinery recovered quickly and was adding capacity in the form of new stacks by 1916.

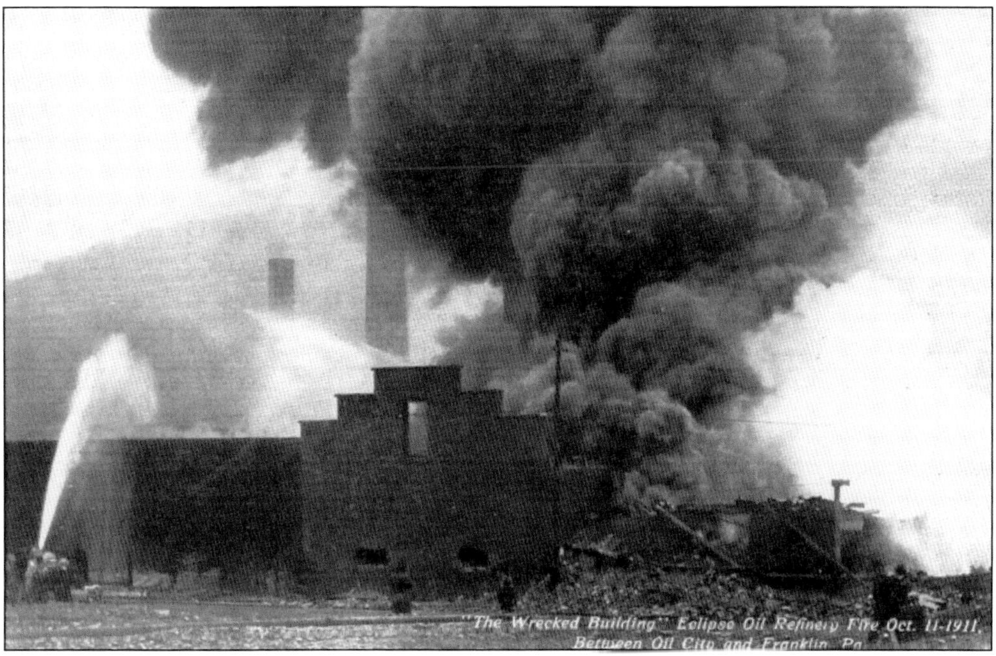

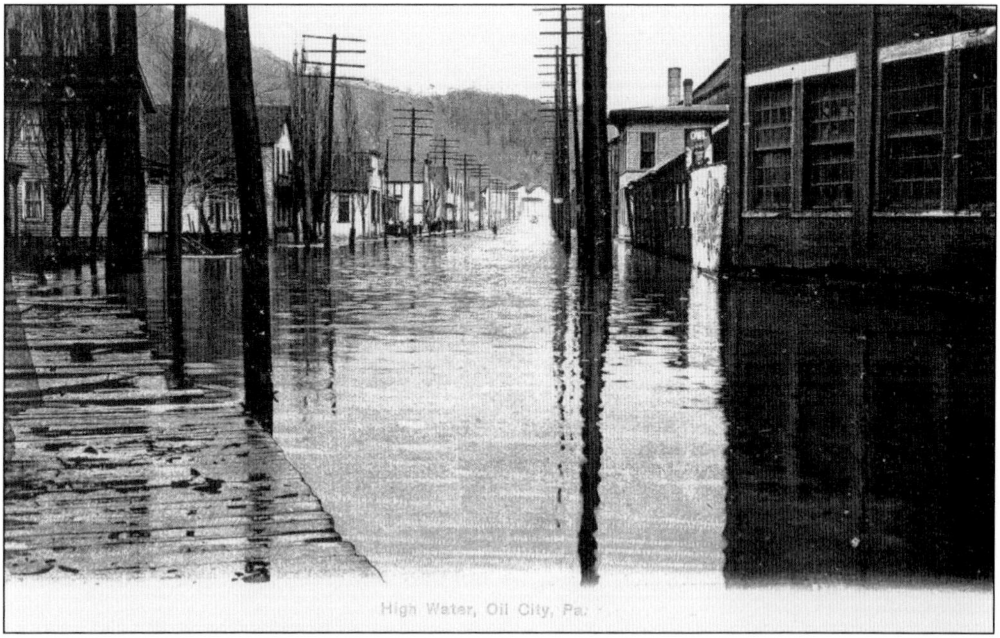

Most towns of the oil region that were built in low-lying areas, especially along Oil Creek and the Allegheny River, were susceptible to flooding in spring and during extreme storm events. This postcard, postmarked 1911, shows flooding along the length of North Seneca Street in Oil City. Note the plank sidewalk at the left of the image.

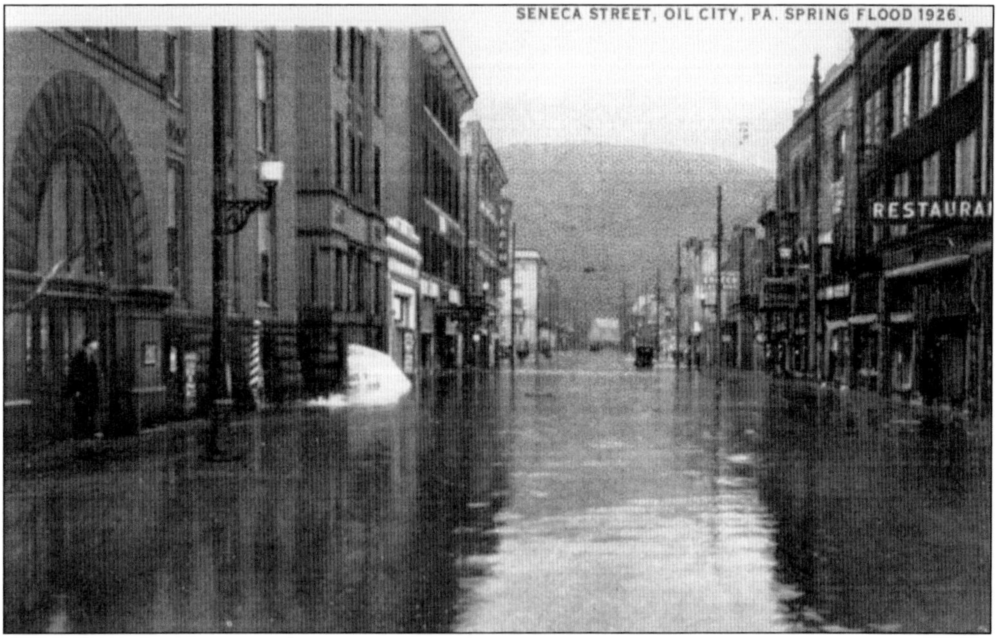

This image shows a flooded Seneca Street on the north side of Oil City, likely due to the spring flood of 1926. The building at the left of the image is the National Transit Building, built in 1890. National Transit Pump and Machine Company was a Standard Oil Company affiliate that controlled 10,000 miles of pipeline and had tank storage capacity for 55 million gallons of crude oil. Today the building houses art galleries and a civic center.

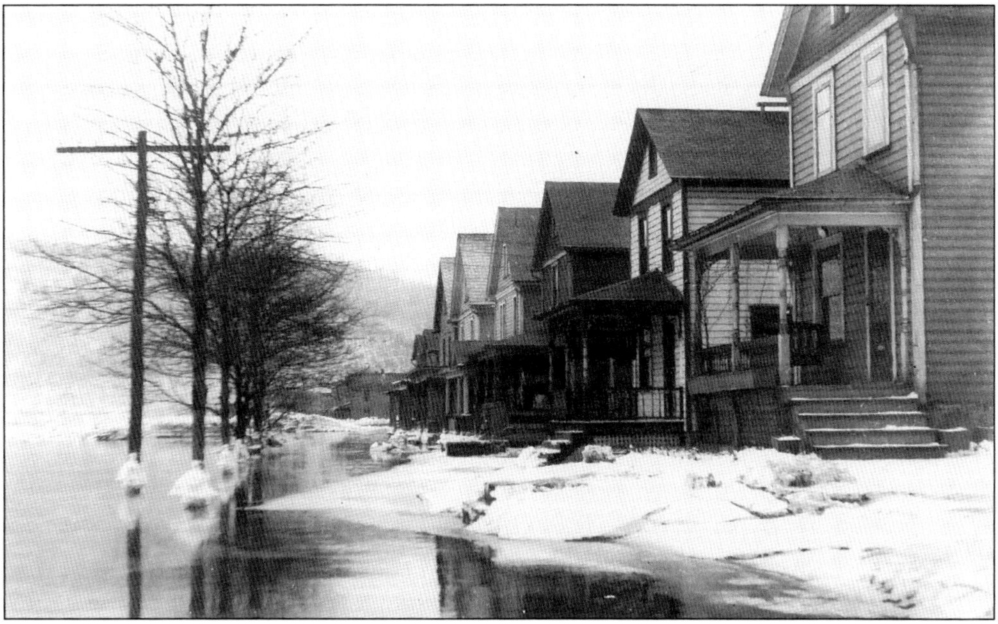

This undated image shows ice jam flooding along the Allegheny River in Franklin, likely during the flood of 1936. Record snowfalls and low temperatures occurred in the region from December 1935 to February 1936, and ice accumulated heavily on streams and rivers. When warming came in February, the ice jammed up, causing flooding of low-lying areas.

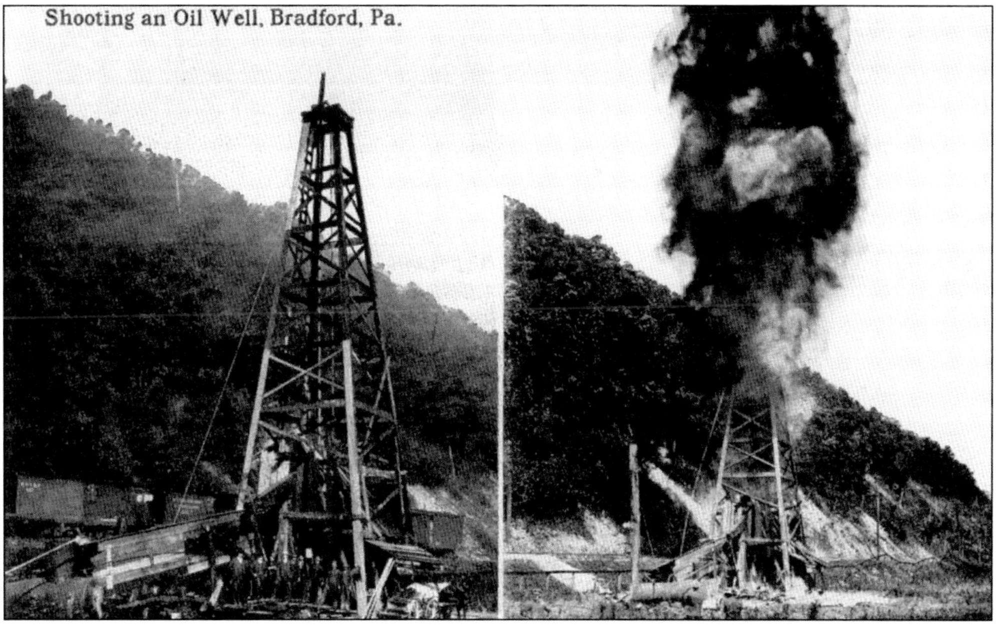

Oil wells sometimes became clogged with paraffin wax and other materials restricting the flow of product. In 1864, Col. E. A. L. Roberts and his brother William developed the torpedo, an explosive charge sent down drill holes of unproductive wells to unclog them and stimulate oil flow. In January 1865, Roberts gave a successful public demonstration of his torpedo on the Ladies Well near Titusville, improving its production. This popular postcard, dated 1913, shows before and after images of well "shooting."

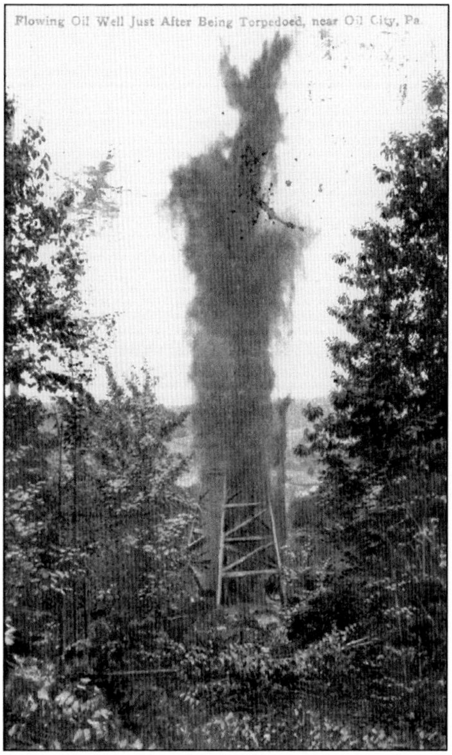

Unlike William Reed's gunpowder-powered torpedo of 1863, the Roberts torpedo eventually used nitroglycerine as an explosive. The Roberts Petroleum Torpedo Company was organized in 1865, and Col. E. A. L. Roberts received a patent on his torpedo in 1866. A "torpedo war" followed, in which oil producers railed against Roberts's monopoly by shooting wells at night, avoiding his fees. Roberts's agents gathered evidence on these "moonlighters," causing hundreds of lawsuits. Roberts's patent was upheld by the Supreme Court in 1880, ending the "war."

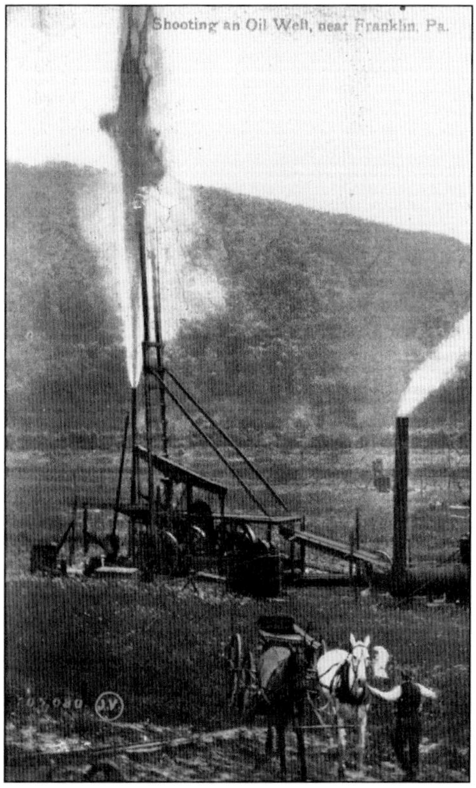

Well shooting became an established profession in the oil region with many practitioners. A well shooter traveled in a specially built "nitro wagon," in which the explosive was stored under the seat. A steady, dependable team of horses was also vital for safety and success. This image from the early 1900s shows a well shooter with his nitro wagon in action somewhere near Franklin.

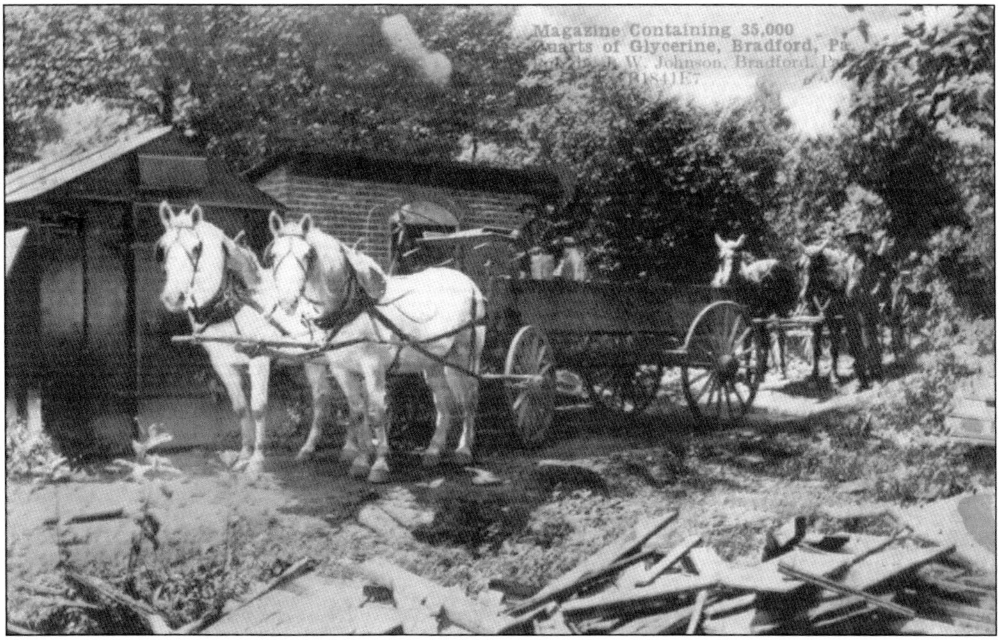

Nitroglycerine was an important, but dangerous, commodity for well shooters in the oil region, who used torpedoes charged with the explosive to stimulate production of stagnant oil wells. Missteps with the explosive caused 31 deaths alone in Venango County, including shooters, employees of wells, and bystanders. Explosions at storage magazines and torpedo factories also caused many deaths. This postcard form the early 1900s shows a "Magazine Containing 35,000 Quarts of Glycerine, Bradford, Pa."

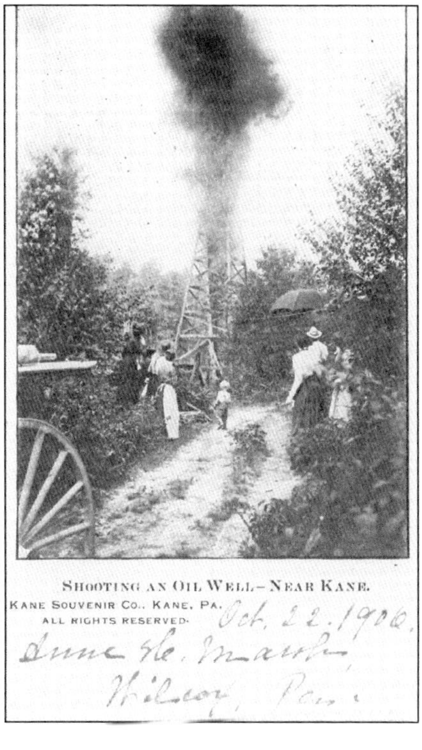

Well shooting was a spectacle and often a source of entertainment in the oil region. It was not unusual for bystanders to a gather about a well shooting or for hosts to entertain visiting dignitaries with a well shoot. This postcard, dated 1906, shows a well being shot near Kane. Note the onlookers some distance from the well, including a young child and a man sporting an umbrella for protection against oil, water, and other flying materials.

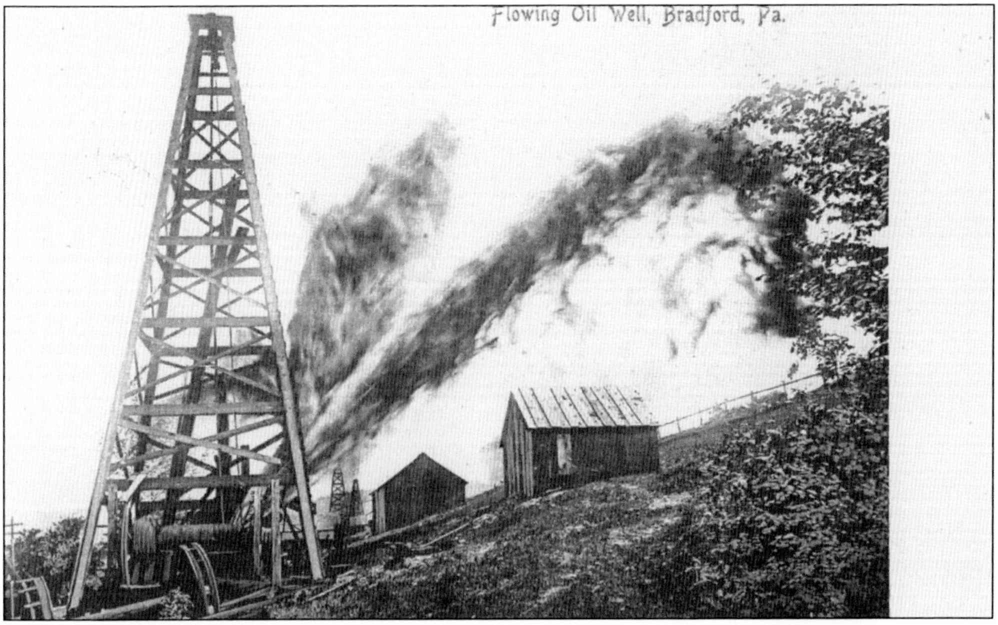

Newly drilled oil wells in the oil region were often placed into two categories: free-flowing and pumping. Free-flowing wells were like artesian wells, in which oil welled up to the surface, sometimes at great pressure, on its own. Pumping wells needed an oil-lifting device, such as a powered pump jack to pull the oil from the ground. This image of a flowing oil well was postmarked in 1906.

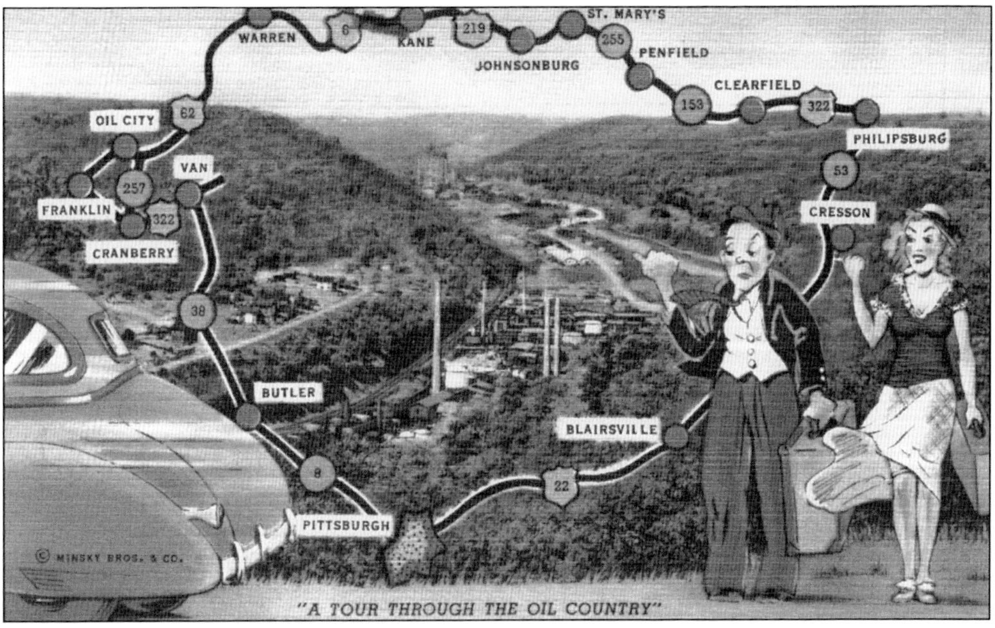

This linen-era postcard sheds a whimsical light on the oil region of western Pennsylvania. Oil Creek and the Pennzoil Refinery at Rouseville are in the background of the image. Although the "Tour Through the Oil Country" mapped on this postcard includes many important areas in the oil region, it also includes many towns outside of the region. No doubt it made for a scenic and interesting tour.

Bibliography

Beers, F. W. *Atlas of the Oil Region of Pennsylvania.* New York: F. W. Beers, A. D. Ellis, and G. G. Soule, Publishers, 1865.

Black, Brian. *Petrolia: The Landscape of America's First Oil Boom.* Baltimore: Johns Hopkins University Press, 2000.

Commonwealth of Pennsylvania, Department of Health. *Report on the Sanitary Survey of the Allegheny River Basin.* Harrisburg, Pennsylvania: W. Stanley Ray, State Printer, 1915.

Cone, Andrew, and Walter R. Johns. *Petrolia: A Brief History of the Pennsylvania Petroleum Region.* New York: D. Appleton and Company, 1870.

Eaton, S. J. M. *A History of the Oil Region of Venango County, Pennsylvania.* Philadelphia: J. P. Skelly and Company, 1866.

Egle, William H. *History of the Commonwealth of Pennsylvania.* Philadelphia: E. M. Gardner, 1883.

Giddens, Paul H. *Pennsylvania Petroleum 1750–1872: A Document History.* Harrisburg: Pennsylvania Historical and Museum Commission, 1947.

Giddens, Paul H. *The Early Days of Oil.* Princeton, New Jersey: Princeton University Press, 1948.

McLaurin, John J. *Sketches in Crude Oil: Some Accidents and Incidents of Petroleum Development in All Parts of the Globe.* Harrisburg: self-published, 1898.

Michener, Carolee K. *Oil, Oil, OIL!* Franklin, Pennsylvania: The Venango County Historical Society, 1997.

Ross, Philip W. *Allegheny Oil: The Historic Petroleum Industry on the Allegheny National Forest.* Warren, Pennsylvania: USDA Forest Service, Allegheny National Forest, 1996.

Williams, Oliver P. *County Courthouses of Pennsylvania: A Guide.* Mechanicsburg, Pennsylvania: Stackpole Books, 2001.

Works Progress Administration. *Pennsylvania: A Guide to the Keystone State.* New York: Oxford University Press, 1940.

Yergin, Daniel. *The Prize: The Epic Quest for Oil, Money, and Power.* New York: Simon and Schuster, 1991.